WITHDRAWN
WRIGHT STATE UNIVERSITY LIBRARIES

**The Role of Evidence in
Risk Characterization**

*Edited by
Peter M. Wiedemann and
Holger Schütz*

Related Titles

Schütz, H., Wiedemann, P. M., Hennings, W., Mertens, J., Clauberg, M.

Comparative Risk Assessment

Concepts, Problems and Applications

2006
ISBN: 978-3-527-31667-0

Bhagwati, K.

Managing Safety

A Guide for Executives

2006
ISBN: 978-3-527-31583-3

Carroll, R.

Risk Management Handbook for Health Care Organizations

3 Volume Set

2006
ISBN: 978-0-7879-8792-3

The Role of Evidence in Risk Characterization

Making Sense of Conflicting Data

Edited by
Peter M. Wiedemann and Holger Schütz

WILEY-VCH Verlag GmbH & Co. KGaA

The Editors

Dr. Peter M. Wiedemann
Research Centre Jülich
Institute of Neurosciences and Biophysics
Programme Group Humans, Environment,
Technology
52425 Jülich
Germany

Dipl.-Pädagog. Holger Schütz
Research Centre Jülich
Institute of Neurosciences and Biophysics
Programme Group Humans, Environment,
Technology
52425 Jülich
Germany

All books published by Wiley-VCH are carefully produced. Nevertheless, authors, editors, and publisher do not warrant the information contained in these books, including this book, to be free of errors. Readers are advised to keep in mind that statements, data, illustrations, procedural details or other items may inadvertently be inaccurate.

Library of Congress Card No.: applied for

British Library Cataloguing-in-Publication Data
A catalogue record for this book is available from the British Library.

Bibliographic information published by the Deutsche Nationalbibliothek
Die Deutsche Nationalbibliothek lists this publication in the Deutsche Nationalbibliografie; detailed bibliographic data are available in the Internet at http://dnb.d-nb.de.

© 2008 WILEY-VCH Verlag GmbH & Co. KGaA, Weinheim

All rights reserved (including those of translation into other languages). No part of this book may be reproduced in any form – by photoprinting, microfilm, or any other means – nor transmitted or translated into a machine language without written permission from the publishers. Registered names, trademarks, etc. used in this book, even when not specifically marked as such, are not to be considered unprotected by law.

Typesetting Thomson Digital, Noida, India
Printing betz-druck GmbH, Darmstadt
Binding Litges & Dopf Buchbinderei GmbH, Heppenheim

Printed in the Federal Republic of Germany
Printed on acid-free paper

ISBN: 978-3-527-32048-6

Contents

Preface *XIII*
Foreword *XV*
List of Contributors *XVII*

1 **Introduction** *1*
Holger Schütz and Peter M. Wiedemann
References *8*

I **From Scientific Analysis to Risk Policy** *11*

2 **Risk Assessment and Risk Communication for Electromagnetic Fields: A World Health Organization Perspective** *13*
T. Emilie van Deventer and Kenneth R. Foster
2.1 Introduction *13*
2.2 Conceptual Framework for Risk Assessment *14*
2.3 EHC on EMFs *14*
2.3.1 ELF Fields *16*
2.3.2 Static Fields *17*
2.4 Comparison Between the WHO ELF-EHC and the California Report *18*
2.5 Communicating about Risks of EMFs *19*
2.6 Discussion *20*
Appendix 1 *22*
References *23*

3 **Characterizing Evidence and Policy Making** *25*
Evi Vogel and Ginevra Delfini
3.1 Introduction *25*
3.2 Science-based Evidence *25*

3.2.1	From Basic Research to Recommendations	26
3.2.2	Officially Appointed Expert Committees and Self-appointed Experts	28
3.2.3	Communication of Recommendations	28
3.3	Society-based Evidence	29
3.4	Policy Making	30
3.4.1	Role of Policy Drafters in Policy Making	30
3.4.2	Role of Politicians in Policy Making	30
3.4.3	Policy Making and the Media	31
3.4.4	Policies	31
3.5	Conclusions	33
	References	34

II Making Sense of Conflicting Data: Evidence Characterization in Different Research Areas 35

4 Basic Principles and Evidence Characterization of the Data from Genetox Investigations 37
Günter Obe and Vijayalaxmi

4.1	Introduction	37
4.2	Cell Cycle	37
4.3	Test Systems	38
4.3.1	COMET Assay to Evaluate Primary DNA Damage	38
4.3.2	Chromosomal Aberrations	39
4.3.3	Micronuclei	41
4.3.4	Sister Chromatid Exchanges	41
4.3.5	Other Assay Systems and Endpoints	42
4.4	Methodological Aspects	42
4.4.1	*In Vitro* Studies	43
4.4.1.1	CAs in HPLs	43
4.4.1.2	CAs in Fibroblasts	43
4.4.1.3	MN	43
4.4.1.4	SCEs	44
4.4.1.5	Metabolic Activation	44
4.4.2	*In Vivo* Studies	44
4.4.2.1	Mammals	44
4.4.2.2	Humans	44
4.5	GLP	45
4.6	Evidence Characterization and Interpretation of Genetox Results	47
4.6.1	Interpretation of Data from One Endpoint	47
4.6.2	Interpretation of Data from Four Endpoints	48
4.6.3	Interpretation of Data from Three Endpoints	48
4.6.4	Interpretation of Data from Two Endpoints	49

4.7	Genetox Studies with Electromagnetic Fields	50
	References	51

5 Animal Studies 55
Alexander Lerchl

5.1	Introduction	55
5.2	Exposure Systems	57
5.3	Sham Exposure and Cage Controls	58
5.4	Replication Studies	59
5.5	Interpretation of Results	60
5.6	Conclusions	60
	References	64

6 Epidemiology 67
Joachim Schüz

6.1	Introduction	67
6.2	Study Types and Risk Estimation	67
6.3	Making Sense of Conflicting Results	71
6.3.1	Temporal Relation Consistent with Cause and Effect	71
6.3.2	Strength of the Association	73
6.3.3	Dose–Response Relationship	73
6.3.4	Consistency Within and Across Studies	74
6.3.5	Specificity	74
6.3.6	Absence of Bias and Confounding	75
6.3.7	Biological Plausibility	76
6.4	Conclusions	77
	References	78

7 Principles and Practice of Evidence Characterization in Environmental Clinical Case Studies 81
Caroline Herr and Thomas Eikmann

7.1	Clinical Environmental Medicine	81
7.2	Assessment of Health Complaints	81
7.2.1	Environmental Attribution	82
7.2.2	Case History	83
7.3	Exposure Assessment and Evaluation	83
7.3.1	Biomonitoring	83
7.3.2	Effect and Susceptibility Monitoring	84
7.3.3	On-site Inspection	84
7.3.4	Ambient Monitoring	84
7.4	Interdisciplinary Clinical Diagnostics	85
7.4.1	Evaluation of Clinical Cases	86
7.5	Conclusions	88
	References	89

III	**Making Sense of Conflicting Data: Procedures for Characterizing Evidence** *91*
8	**Characterizing Evidence with Evidence-based Medicine** *93*
	Alexander Katalinic and Dagmar Lühmann
8.1	What is Evidence-based Medicine? *93*
8.2	EbM Process *94*
8.3	Five Steps of EbM *94*
8.3.1	Asking Answerable Questions *95*
8.3.2	Finding the Best Available Evidence *95*
8.3.3	Critical Appraisal *97*
8.3.4	Acting on the Evidence *98*
8.3.5	Evaluate your Performance *98*
8.4	Comparing the EbM to Other Approaches of Characterizing Evidence *99*
	References *99*
9	**The *IARC Monographs*' Approach to Characterizing Evidence** *101*
	Vincent James Cogliano, Robert Alexander Baan, Kurt Straif, Yann Grosse, Marie Béatrice Secretan, and Fatiha El Ghissassi
9.1	Introduction *101*
9.2	Pertinent Data for Carcinogen Identification *101*
9.3	International Agency for Research on Cancer Evaluations *102*
9.3.1	Evaluating Epidemiologic Studies *102*
9.3.2	Evaluating Bioassays in Experimental Animals *104*
9.3.3	Evaluating Mechanistic and Other Relevant Data *105*
9.3.4	Overall Evaluation *105*
9.4	Hazard versus Risk *107*
9.5	Ensuring Impartial Evaluations *108*
9.6	Characterizing Evidence in the Future *108*
	References *109*
10	**The Swiss Health Risk Approach** *111*
	Martin Röösli
10.1	Background *111*
10.2	Aims *112*
10.3	Approach *112*
10.3.1	Evidence Rating *112*
10.3.2	Relevance to Health *113*
10.3.3	Exposure Levels *113*
10.3.4	Summary Scheme *113*
10.4	Discussion *116*
10.4.1	Gradual Rating of the Evidence *116*
10.4.2	Source-specific Evaluation *116*
10.4.3	Lack of Data *117*
10.4.4	Publication Bias *117*

10.4.5	Rating of the Study Quality	*118*
10.4.6	Meta-analyses	*118*
10.5	Conclusions	*119*
	References	*119*

11 Procedures for Characterizing Evidence: German Commission on Radiation Protection (Strahlenschutzkommission) *121*
Norbert Leitgeb

11.1	Introduction	*121*
11.2	Assessment of Scientific Evidence	*122*
11.3	Relevance to Human Health	*123*
11.4	Weight of Evidence	*124*
11.5	Multidisciplinary Assessment	*125*
11.6	Regulations	*126*
11.7	Precautions	*126*
11.8	Electromagnetic Interference	*128*
11.9	Conclusions	*128*
	References	*128*

12 Lessons from the California Electromagnetic Field Risk Assessment of 2002 *131*
Raymond Richard Neutra

12.1	Introduction	*131*
12.2	Policy Questions and Questions about Causal Claims and Arguments	*131*
12.3	Bradford Hill's and Koch's Questions	*132*
12.4	The Asymmetry of Some "Rule In" Tests	*138*
12.5	Toulmin's Argument Anatomy and Bayes' Theorem as a Universal Warrant	*139*
12.6	Special Importance of Coherence	*142*
12.7	Plausibility, Experimentation and Analogy	*143*
12.8	Causal Arguments Can Become More Transparent but Will Always Involve Judgment	*147*
	References	*149*

13 Evidence Maps – A Tool for Summarizing and Communicating Evidence in Risk Assessment *151*
Holger Schütz, Peter M. Wiedemann, and Albena Spangenberg

13.1	Introduction	*151*
13.2	Evidence Maps Approach	*151*
13.2.1	Background	*151*
13.2.2	Structure of Evidence Maps	*153*
13.2.3	Constructing an Evidence Map: Cancer Epidemiology	*154*
13.3	Insights from the Process of Building Evidence Maps	*158*
13.4	Conclusions	*159*
	References	*159*

IV Psychological and Ethical Aspects in Dealing with Conflicting Data and Uncertainty *161*

14 Perception of Uncertainty and Communication about unclear Risks *163*
Peter Wiedemann, Holger Schütz, and Andrea Thalmann

14.1 Introduction *163*
14.2 Uncertainty in Risk Assessment *164*
14.3 Uncertainty Communication and Lay Persons' Perception of Uncertainty Information *166*
14.3.1 Intuitive Toxicology: How do Nonexperts Understand the Risk Assessment Framework? *166*
14.3.2 How do People Understand Information about Relative Risks *168*
14.3.3 Information about Uncertainty in Risk Assessment: How do Nonexperts Cope With It? *169*
14.3.4 Uncertainty Descriptions: How do People Understand Qualitative, Quantitative and Visual Expression? *171*
14.3.5 Contextual Effects *174*
14.4 Lay Peoples' Perception of Precautionary Measures *176*
14.5 Outlook and Conclusions *178*
References *179*

15 Ethical Guidance for Dealing with Unclear Risk *185*
Armin Grunwald

15.1 Ethical Guidance in Cases of Unclear Risk – The Challenge *185*
15.2 Entry Points of Ethical Reflection in Situations of Unclear Risk *186*
15.2.1 Entry Points of Ethical Reflection in General *186*
15.2.2 Unclear Risk: Nonstandard Situations with Respect to Risk *187*
15.2.3 Moral Conflicts in Situations of Unclear Risk *189*
15.3 Ethical Approaches to (Unclear) Risk *191*
15.3.1 Consequentialist Approach *191*
15.3.2 Principle of Pragmatic Consistency *192*
15.3.3 "Imperative of Responsibility" (Jonas) *193*
15.3.4 Projected Time *194*
15.3.5 Deontological Advice *195*
15.3.6 Interim Conclusions (1) *196*
15.4 Operative Approaches *197*
15.4.1 Precautionary Principle *197*
15.4.2 Principle of Prudent Avoidance *198*
15.4.3 Interim Conclusions (2) *199*
15.5 Conclusions *200*
References *201*

V	**Practical Implications** *203*	

16	**Lessons Learned: Recommendations for Communicating Conflicting Evidence for Risk Characterization** *205*	
	Peter Wiedemann, Franziska Börner, and Holger Schütz	

16.1 Introduction *205*
16.2 Guiding Principles in Risk Communication *207*
16.2.1 Prudence *207*
16.2.1.1 Assess the Underlying Problem *207*
16.2.1.2 Both Content and Process do Matter *207*
16.2.2 Transparency *208*
16.2.2.1 Make Your Expertise Transparent *208*
16.2.2.2 Describe the Context of Your Work and the Process of Arriving at the Conclusion *208*
16.2.2.3 Reveal your Evaluation Framework *208*
16.2.2.4 Describe the Rules that You Use for Evaluating the Weight of Evidence *209*
16.2.3 Impartiality *209*
16.2.3.1 Give the Pros and Cons of Your Assessment *209*
16.2.3.2 Depict the Remaining Uncertainties but Do Not Forget to Point Out the Evidence Already Available *209*
16.2.4 Reasonableness *210*
16.2.4.1 Explain the Process of Evaluating Evidence *210*
16.2.4.2 Explain the Relevance of the Endpoints for Evaluating Human Health Risks *210*
16.2.4.3 Put the Available Evidence in Perspective *210*
16.2.4.4 Support Accessibility of Critical Information *211*
16.2.4.5 Assess the Potential Risk *211*
16.2.4.6 Put the Potential Risk in Perspective *211*
16.2.5 Clarity *211*
16.2.5.1 Give No More Information than Necessary *211*
16.2.5.2 Be Aware of Your Language *211*
16.2.5.3 Test the Perceptions of your Communication Formats *212*
16.2.6 Responsibility *212*
16.2.6.1 How Much Evidence is Evidence Enough for Taking Action? *212*
References *213*

Index *215*

Preface

The problem of characterizing and summarizing evidence within risk assessment became noticeable to us working in two research projects. Both focused on the evidence evaluation for potential health risks of radiofrequency (RF) electromagnetic fields (EMFs) from mobile communication. The projects not only provided the opportunity to study how researchers within particular research fields evaluate evidence, they also provided an insight on how experts handle the problem of conflicting data. These research opportunities helped us to develop an understanding of how evidence from different research areas can be integrated into an overall risk assessment. Additionally, they also gave us valuable insights about the various procedures that are used by research groups or institutions for communicating the results of their risk assessments.

These experiences and insights motivated us to conduct a workshop with the title "Characterizing Evidence in EMF Risk Assessment" in May 2006 in Berlin, Germany. International researchers from as different research fields as genotoxicity testing, epidemiology, public health and risk communication discussed several procedures that are presently used by national or international agencies and other organizations to evaluate and communicate what is known about the potential health risks from extremely low-frequency and/or RF EMFs. Inspired by the lively workshop discussions, highlighting the complexity and variety of approaches, the idea was conceived to publish a book on evidence characterization in risk assessment.

A further impetus for the book came from an ongoing research project sponsored by the German Helmholtz Research Association on the Implications of Biomedicine for the Assessment of Human Health Risks (IMBA). The research focus of this project is how specific new developments in biomedicine summarized under the term "toxicogenomics" will transform the present risk management framework. Again, a central issue is to characterize the evidence about health risks from RF EMFs. The IMBA project, however, tackles the issue of "evidence characterization" from two perspectives. First, the project focuses specifically on the existing knowledge gaps in RF EMF cancer risk assessment. These knowledge gaps – which are not due to missing, but to conflicting data – provide the starting point for the IMBA project. Second, the problem of how to characterize evidence emerges with respect

The Role of Evidence in Risk Characterization: Making Sense of Conflicting Data.
Edited by Peter M. Wiedemann and Holger Schütz
Copyright © 2008 WILEY-VCH Verlag GmbH & Co. KGaA, Weinheim
ISBN: 978-3-527-32048-6

to the "omics" studies on RF EMFs itself. Here, a key issue is to determine which gene expressions can provide reliable and valid indicators of toxicity at earlier stages and at lower doses than traditional toxicology parameters.

New scientific developments are likely to reveal new problems for evidence characterization, but with the present book we hope to provide an overview of the current state of knowledge on how to deal with conflicting data and to evaluate evidence in risk characterization.

We would like to thank T-Mobile, Germany, who sponsored both research projects on potential health risks of RF EMFs from mobile communication and who partly sponsored the Berlin workshop. We are also grateful to the Helmholtz Association of German Research Centers that supports the IMBA project.

November 2007

Peter Wiedemann
Holger Schütz
Jülich
Germany

Foreword

In recent years, increasing concern has been expressed about the possible long-term adverse consequences of new technologies, which have been developing and spreading into the market much more rapidly than the time needed for scientific research to investigate their impact on human health and the environment. The problem therefore arises how to implement protection measures that may prevent hypothesized, but not proven, effects and at the same time are adequately supported by scientific findings: in other words, how to balance science and precaution.

This is the core of present, lively debates on the precautionary principle and related health policies, and electromagnetic fields (EMF) represent a paradigmatic case study in this regard.

Studies on biological and health effects of both low- and high-frequency EMFs have been conducted for more than 50 years, and the number of peer-reviewed papers amounts to several thousands. Based on this scientific database, some international organizations have developed exposure standards that have evolved over time from simple recommendations for limitation of exposure to a few specific sources, to complex and sophisticated protection systems that encompass all the nonionizing range of the electromagnetic spectrum.

The International Commission on Non-Ionizing Radiation Protection (ICNIRP), an independent scientific body officially recognized by the World Health Organization (WHO), in 1998 issued guidelines for the safe exposure of workers and the general public to EMFs in the frequency range from 0 Hz (static fields) to 300 GHz (upper limit of microwaves). Exposure limits recommended by the ICNIRP provide adequate protection against all *established* adverse effects of exposure. Effects are considered as established if they are indicated by high-quality studies, are replicable and are consistent with findings from other research areas (e.g. results obtained *in vivo* are biologically coherent consistent with results *in vitro*).

All health effects of EMF established so far are acute in nature and exhibit thresholds, in the sense that they are only detectable when the exposure level (in terms of appropriate biologically effective quantities) exceeds a given value. Exposure limits set in the ICNIRP guidelines – as well as those developed by other internationally recognized bodies – are well below the thresholds, thanks to the conservative hypotheses adopted and to the introduction of relevant reduction factors.

The Role of Evidence in Risk Characterization: Making Sense of Conflicting Data.
Edited by Peter M. Wiedemann and Holger Schütz
Copyright © 2008 WILEY-VCH Verlag GmbH & Co. KGaA, Weinheim
ISBN: 978-3-527-32048-6

While international standards are only based on established, short-term effects, increasing concern has been expressed about the possible development of long-term degenerative pathologies as a consequence of chronic exposure to EMFs, although below lower than the limits. At the same time, an increasing number of people claim to suffer subjective symptoms that they attribute to EMFs, although no convincing evidence of a causal relationship has been found in several controlled studies. As a result of these concerns, strong pressure has been exerted by the public on national and local authorities for the implementation – in the name of the precautionary principle – of protection measures more stringent than the international standards.

The precautionary principle establishes that preventive actions should be needed whenever some evidence exists, although uncertain and incomplete, that an agent, a substance or a technology may constitute a hazard for human health or the environment. The conditions under which the principle is applicable and the way related measures should be implemented have been widely discussed. In particular, it has been remarked that, for the principle to be invoked, a potentially serious hazard must have been identified and scientifically evaluated. In addition, measures taken in the name of the precautionary principle should in proportion to the risk they aim at protecting and this in turn requires some quantitative estimate of the risk itself. In conclusion, scientific evidence is a key requisite not only for science-based exposure guidelines, but for any precautionary policy.

Although the scientific database is the same, a variety of approaches have been adopted to the development of precautionary policies. This is due to different social attitudes and cultures, but also to different consideration for gaps and uncertainties in scientific knowledge. Policies implemented in various countries show large differences from one another and also internal inconsistencies. For example, in many cases the same level of precaution is adopted with respect to low-frequency sources (such as power lines) and high-frequency sources (such as mobile phones and their base stations) notwithstanding that the evidence of health risks is not the same: while extremely low-frequency magnetic fields have been classified as *possibly carcinogenic to humans* by the International Agency for Research on Cancer, studies on radiofrequency EMFs do not suggest, overall, adverse health effects for exposures below international limits.

The aim of this book is to show how evidence characterization is done in different research areas, to discuss different approaches to evidence characterization, and to explore psychological and ethical aspects relevant for evidence characterization as a first step toward a better understanding of scientific data, more effective risk communication and possibly harmonization of risk policies.

Paolo Vecchia
National Institute of Health
Rome
Italy

List of Contributors

Robert Alexander Baan
International Agency for Research on Cancer (IARC)
150 Cours Albert Thomas
69372 Lyon CEDEX 08
France

Franziska Börner
Research Centre Jülich
Institute of Neurosciences and Biophysics
Programme Group Humans, Environment, Technology
52425 Jülich
Germany

Vincent James Cogliano
International Agency for Research on Cancer (IARC)
150 Cours Albert Thomas
69372 Lyon CEDEX 08
France

Ginevra Delfini
Dutch Ministry of Housing, Spatial Planning and the Environment
Radiation Protection, Nuclear and Biosafety Division
P.O. Box 30945
2500 GX The Hague
The Netherlands

Thomas Eikmann
Universitätsklinikum Giessen und Marburg GmbH
Institut für Hygiene und Umweltmedizin
Friedrichstr. 16
35392 Giessen
Germany

Fatiha El Ghissassi
International Agency for Research on Cancer (IARC)
150 Cours Albert Thomas
69372 Lyon CEDEX 08
France

Kenneth R. Foster
University of Pennsylvania
Department of Bioengineering
240 Skirkanich Hall
210 S. 33rd Street
Philadelphia PA 19104-6392
USA

Yann Grosse
International Agency for Research on Cancer (IARC)
150 Cours Albert Thomas
69372 Lyon CEDEX 08
France

The Role of Evidence in Risk Characterization: Making Sense of Conflicting Data.
Edited by Peter M. Wiedemann and Holger Schütz
Copyright © 2008 WILEY-VCH Verlag GmbH & Co. KGaA, Weinheim
ISBN: 978-3-527-32048-6

Armin Grunwald
Forschungszentrum Karlsruhe
Institut für Technikfolgenabschätzung
und Systemanalyse (ITAS)
Hermann von Helmholtz Platz 1
76344 Eggenstein-Leopoldshafen
Germany

Caroline Herr
Bayerisches Landesamt für Gesundheit
und Labensmittelsicherheit
Veterinärstrasse 2
85764 Oberschleißheim
Germany

Alexander Katalinic
Universität Lübeck
Institut für Krebsepidemiologie e.v.
Beckergrube 43–47
23552 Lübeck
Germany

Norbert Leitgeb
Graz University of Technology
Institute of Health Care Engineering
Inffeldgasse 18/I
8010 Graz
Austria

Alexander Lerchl
Jacobs University Bremen
School of Engineering and Science
Campus Ring 6
28725 Bremen
Germany

Dagmar Lühmann
Universitätsklinikum Schleswig-
Holstein (Campus Lübeck)
Institut für Sozialmedizin
Beckergrube 43–47
23552 Lübeck
Germany

Raymond Richard Neutra
California Department of Public Health
Division of Environmental and
Occupational Disease Control
Current Address:
956 Evelyn Avenue
Albany, CA 94706
USA

Günter Obe
Universität Duisburg-Essen
Essen
Germany
Current address:
Gershwinstrasse 33
14513 Teltow
Germany

Martin Röösli
University of Bern
Department of Social and Preventive
Medicine
Finkenhubelweg 11
3012 Bern
Switzerland

Holger Schütz
Research Centre Jülich
Institute of Neurosciences and
Biophysics
Programme Group Humans,
Environment, Technology
52425 Jülich
Germany

Joachim Schüz
Danish Cancer Society
Institute of Cancer Epidemiology
Strandboulevarden 49
2100 Copenhagen
Denmark

Marie Béatrice Secretan
International Agency for Research on
Cancer (IARC)
150 Cours Albert Thomas
69372 Lyon CEDEX 08
France

Albena Spangenberg
Research Centre Jülich
Institute of Neurosciences and
Biophysics
Programme Group Humans,
Environment, Technology
52425 Jülich
Germany

Kurt Straif
International Agency for Research on
Cancer (IARC)
150 Cours Albert Thomas
69372 Lyon CEDEX 08
France

Andrea T. Thalmann
T-Mobile Deutschland GmbH
Information and Sustainability
Eschollbrücker Straße 12
64283 Darmstadt
Germany

T. Emilie van Deventer
World Health Organization
Radiation & Environmental Health
Public Health and Environment
21 Avenue Appia
1211 Geneva 27
Switzerland

Paolo Vecchia
National Institute of Health
Department of Technology and Health
Viale Regina Elena, 299
00161 Rome
Italy

Vijayalaxmi
University of Texas
Health Science Center
Department of Radiation Oncology
San Antonio
Texas
USA

Evi Vogel
Bayerisches Staatsministerium für
Umwelt, Gesundheit und
Verbraucherschutz
Postfach 810140
81901 München
Germany

Peter M. Wiedemann
Research Centre Jülich
Institute of Neurosciences and
Biophysics
Programme Group Humans,
Environment, Technology
52425 Jülich
Germany

1
Introduction

Holger Schütz and Peter M. Wiedemann

Risk characterization is a fundamental task in risk assessment. It constitutes the final step in the process of risk assessment, which starts with hazard identification and is followed by dose–response assessment and exposure assessment [1, 2].

> Risk characterization is the summarizing step of risk assessment. The risk characterization integrates information from the preceding components of the risk assessment and synthesizes an overall conclusion about risk that is complete, informative and useful for decision makers. ([3], p. A5)

Without a qualified summary of the evidence that highlights the essential and critical insights of the risk assessment, its results will remain useless for risk communication and risk management.

In risk assessment the evidence with regard to a causal relationship between exposure to a particular agent and an adverse health outcome is crucial, because it determines whether a hazard exists (hazard identification). How hazard identification is actually performed and on which evidence it is based depends, of course, on the type of hazard. Usually, however, hazard identification requires evidence from different research areas; evidence from one research field alone, let alone findings from single studies, is not sufficient. The International Agency for Research on Cancer (IARC), for instance, considers evidence from epidemiology, animal research and *in vitro* studies for its evaluation of the carcinogenicity of agents (see Ref. [4] and Chapter 9 of this volume).

Summarizing and integrating evidence from different research fields, such as epidemiology or animal studies, into an overall evaluation regarding the hazard potential of an agent is a demanding task. A particular problem arises when the evidence consists of a multitude of inconsistent or even contradictory results from scientific studies. The problem is increased if not only a single, but several adverse health effects are involved, each having its own set of contradictory research findings. This is a situation not unfamiliar to risk assessment experts, but it can be confusing for nonexperts. It is certainly not surprising that learning about conflicting evidence regarding a potential adverse health effect irritates the public. In reporting about health risks the media often focus on expert dissent ([5]; for a more general account of

The Role of Evidence in Risk Characterization: Making Sense of Conflicting Data.
Edited by Peter M. Wiedemann and Holger Schütz
Copyright © 2008 WILEY-VCH Verlag GmbH & Co. KGaA, Weinheim
ISBN: 978-3-527-32048-6

what constitutes the value of news in the media, see Ref. [6]) and tend to report the ideas of maverick scientists as representing an accepted opinion in the scientific community, thus conveying the impression of a balanced controversy [7]. With regard to risk communication a central question is whether – and if how – uncertainties in risk assessment, for which expert dissent is a major source, should be communicated to the public.

Conflicting evidence is barely less irritating for risk regulators who need conclusive evidence about a potential hazard to justify their regulations. Typically, regulating a particular technology or a specific substance requires a "proof" for hazardousness, i.e. requires undisputable scientific evidence that using the technology or being exposed to the substance actually causes an adverse health effect. In recent years, however, there have been increasing demands to relax that strict requirement and to act according to the precautionary principle. Essentially, the precautionary principle states (in the words of the Wingspread Statement on the Precautionary Principle) that "[w]here an activity raises threats of harm to the environment or human health, precautionary measures should be taken even if some cause and effect relationships are not fully established scientifically " ([8], p. xiii). The precautionary principle is not uncontroversial as guidance for risk-related policy making [9–11]; however, the fact that it plays such an important role in recent debates about regulating potential hazards in situations with conflicting evidence points to the significance of thinking about the normative rules that can be used for decision making under uncertainty.

One can easily provide a long list of issues that have been scientifically controversial as to whether they bear a human health risk or not and for which claims have been made that they should be regulated according to the precautionary principle. Well-known examples include genetically modified organisms, endocrine disruptors or synthetic nanoparticles.

Another excellent example – which is used throughout this book – is the potential health risks from electromagnetic fields (EMFs). Extremely low-frequency (ELF) magnetic fields have been a public health issue since 1979 when Wertheimer and Leeper [12] published a study that pointed to a possible link between exposure to 50/60-Hz fields from power lines and childhood leukemia. Since then, quite a number of studies have been conducted, also with regard to other health outcomes, without being able to clearly demonstrate a causal relationship between ELF field exposure and the respective health outcome [13, 14]. However, due to the evidence from epidemiological studies, IARC classified ELF magnetic fields as "possibly carcinogenic to humans" [15]. Despite this classification, potential health threads from ELF magnetic fields received much less public (and media) attention than those from radiofrequency (RF) EMFs. This issue has raised broad public concern in many countries, and is an ongoing topic in public debate and popular media [16]. On the regulatory level the controversy is about whether the existing exposure limits are sufficient to protect public health (these limits have been suggested by the International Commission on Non-Ionizing Radiation Protection (ICNIRP) [17]). These exposure limits are based on the established scientific evidence about the thermal (i.e. heating) effects of RF EMF exposure. However, whether exposures below these limits might cause adverse health effects is scientifically controversial. Over recent years, a number of international expert reports have evaluated the scientific evidence on potential health risks from EMF exposure below the exposure limits [18–23]. Health outcomes that have been considered in these reports include such diverse endpoints as brain cancer, DNA damage, the blood–brain barrier, cognitive performance, sleep, the nervous system or melatonin production. All reports agree that there is no scientific proof of health risks below the exposure limits – they disagree, however, with regard to the extent and the relevance of uncertainties in the scientific evidence. They also disagree whether precautionary measures should be implemented or not. As a result, some countries, such as Switzerland, Poland and Luxembourg, implemented stricter precautionary limit values, while many countries adopted the ICNIRP recommendations for exposure limits.

The preceding outline points to four problem areas that deserve attention when addressing the question of how do deal with conflicting evidence in risk characterization:

(I) How does the scientific analysis of risk inform policy making about potential hazards or risks?
(II) How is evidence characterized within different research areas?
(III) How can evidence from different research areas be integrated into an overall risk characterization?
(IV) How do lay people understand the uncertainties in risk assessment and how can these be communicated to the public – and are there any ethical guidelines that can help in making risk-related decisions under uncertainty?

The book is organized along these questions. Part I (*From Scientific Analysis to Risk Policy*) addresses the question how the scientific analysis of risk informs risk policy.

The probably most influential single organization for the regulation of human health threads is the World Health Organization (WHO). Although countries usually do have their own national scientific bodies for assessing and evaluating hazards and risks, they often base their standard setting regarding human health threads on WHO assessments. Emilie van Deventer and Kenneth Foster describe the process the WHO uses to identify and characterize the evidence regarding potential health risks from static and ELF magnetic fields, and show how this process is complemented by risk communication activities that translate the expert-oriented assessments in a way that is more easily accessible and understandable to a broader lay public.

Evi Vogel and Ginevra Delfini, in their chapter on evidence characterization and policy making, discuss the various aspects that influence risk-related regulatory decision making. They argue that for actual policy making, science-based evidence regarding a hazard is not the only contributing part. Of considerable importance is also what they call "society-based evidence", i.e. the multitude of public reactions regarding the hazard, such as different risk perceptions or interests of particular societal groups (e.g. industry, activist movements). Policy makers cannot ignore these voices when deciding about exposure limits or other types of risk regulations. Vogel and Delfini emphasize furthermore the key role of "policy drafters" in the actual preparation of regulatory decisions. Their task is to integrate all the scientific and societal sources of information, and give each appropriate weight. They are the ones who prepare the actual policies upon which the political decision makers ultimately decide.

The chapters by van Deventer and Foster as well as Vogel and Delfini illustrate that the process from scientific evidence to policy making includes many aspects which go beyond "pure science". Nevertheless, scientific analysis provides the basis for risk assessment. Part II (*Making Sense of Conflicting Data: Evidence Characterization in Different Research Areas*) is therefore dedicated to how evidence characterization actually takes place in those scientific fields that are most important for human health risk assessment: genotoxicity testing, animal studies, epidemiology and clinical case studies.

Günter Obe and Vijayalaxmi present the basic principles and evidence characterization of genetic toxicology, which investigates the adverse effects of exposure to a biological, chemical or physical agent on the DNA. They briefly introduce the test strategies which are usually applied (COMET assay, chromosomal aberrations, micronuclei and sister chromatid exchanges), and discuss the methodological standards and problems associated with their use in *in vitro* and *in vivo* studies. Each of the test strategies allows only limited conclusions, and Obe and Vijayalaxmi discuss how much evidence for the genotoxicity of an agent is given by positive results from one or more test strategies or combinations of positive and negative results. Applying this evaluation strategy to the evidence from genotoxicity studies leads the authors to the conclusion that the available data do not support the notion of genotoxic effects from RF EMFs.

Animal studies are one of the cornerstones of risk assessment. Alexander Lerchl shows that for evaluating the evidence from those experiments, one has to carefully observe the concrete experimental setting that has been used. With regard to animal studies on the effects of RF EMFs on malignancies in rodents – which is the example Lerchl uses to illustrate his points – critical issues are the quality of the exposure system used and the monitoring of the actual exposure of the test animals. Another issue is that the experimental and the control condition must only differ with regard to exposure. This means that the control condition has to be set up as a sham exposure – being an exact copy of the exposure condition except for the exposure itself. Other important methodological requirements include cage controls and blinded designs. Lerchl argues that the inconsistencies regarding carcinogenic effects of RF EMF exposure, where some early animal studies yielded positive results while the more recent studies have been all negative, can be explained – at least in part – through methodological deficits of the early studies.

Both genotoxicity testing and animal studies have the advantage of allowing experimental designs that provide the strongest information regarding a causal relationship between the exposure with an agent and an adverse effect. However, both also share the problem of extrapolating their results to human beings – the real target in risk assessment. Since experiments with humans are for obvious reasons not an option – at least with regard to severe endpoints such as cancer – the only way to directly investigate the adverse effects of agents on humans is through epidemiological studies. The drawback of epidemiological studies, however, is that they do not permit us to directly detect causal relationships. In his chapter on epidemiology, Joachim Schüz examines this problem in detail. He first provides a brief overview of the strength and weaknesses of different study types used in epidemiology. Using the example of epidemiological research on ELF magnetic fields and leukemia, Schüz then discusses the criteria which can be used to evaluate the epidemiological evidence regarding a causal relationship. In applying these criteria to the epidemiological evidence regarding ELF magnetic fields and leukemia, Schüz concludes that there is limited evidence for a causal relationship between ELF magnetic field exposure and childhood leukemia, but that bias, chance or confounding could not be ruled out as alternative explanations.

Genotoxicity testing, animal studies and epidemiology have one common characteristic: they all investigate whether exposure with a specific agent, e.g. RF EMFs, has an adverse impact on a selected endpoint. In environmental clinical case studies the situation is almost reverse. Here, the focus is on the health impairment of an individual patient and the problem is to identify the environmental exposures which are responsible for that. This change in perspective also changes, to some extent, the meaning of evidence characterization, as Caroline Herr and Thomas Eikmann show in their chapter. In environmental clinical case studies the main source for evidence is the patient's individual case history which the physician has to elicit in a structured manner. It includes not only the present health impairment and its development, but also the causes to which the patient attributes their impairment. In this context, evidence characterization means to evaluate whether those environmental exposures which a patient's case history or personal attribution suggest as possible causes of health impairment do indeed provide a plausible explanation.

How can the findings from the different scientific fields discussed in Part II be integrated into an overall scientific picture for risk assessment? Part III (*Making Sense of Conflicting Data: Procedures for Characterizing Evidence*) illustrates this by presenting six procedures for characterizing evidence. Four of these procedures have been specifically developed for assessing potential health risks of ELF and/or RF EMFs. However, it will become evident that all presented procedures can be applied to almost any potential hazard.

At first, Alexander Katalinic and Dagmar Lühmann provide a brief account of evidence-based medicine (EbM). Although usually applied in a different context – healthcare decision making – EbM is paradigmatic for characterizing and summarizing evidence. It consists of a systematic and elaborated sequence of steps, starting with a precise and practicable problem definition, followed by a systematic search for and evaluation of relevant research results. Finally, useful findings are implemented in clinical practice and evaluated with regard to their performance. Of particular interest in the context of evidence characterization for risk assessment is the grading system which is used for evaluating the relevant research results. It distinguishes five levels of evidence, most of them having additional sublevels, which refer to the quality of the study design. In the EbM system, randomized controlled trials yield the highest level of evidence, while uncritical expert opinion is considered to be of low evidential power.

For risk assessment, one of the assuredly best known and most influential procedures to characterizing evidence is the approach which the International Agency for Research on Cancer (IARC) uses in its evaluation of carcinogenic risks to humans. In their chapter, Vincent Cogliano and his colleagues from the IARC describe the elements of this approach. The IARC bases its evaluation of carcinogenicity most importantly on evidence from long-term bioassays in experimental animals and on epidemiological studies, but also considers findings from mechanistic studies and other relevant data. The evidence from animal and epidemiological studies is characterized with standard descriptors, ranging from *sufficient evidence of carcinogenicity* to *evidence suggesting lack of carcinogenicity*. For the overall evaluation, which finally integrates the evaluations of epidemiology, animal and mechanistic studies by using an elaborated set of rules, the IARC uses five groups. An agent would be classified in Group 1: *carcinogenic to humans*, if there was sufficient evidence in humans. Less conclusive evidence would lead to classifications as *probably* (Group 2A) or *possibly* (Group 2B) carcinogenic or even as *not classifiable as to its carcinogenicity to humans* (Group 3). An agent would be classified as *probably not carcinogenic to humans* (Group 4) if evidence from both epidemiological and animal studies was suggesting lack of carcinogenicity.

The Swiss Health Risk Approach, which is outlined by Martin Röösli, uses a level of evidence categorization that is quite similar to the IARC classification scheme. Unlike the IARC approach it is not confined to cancer endpoints, but includes a variety of other health relevant effects. The levels of evidence which are used are *established*, *probable*, *possible*, *improbable* and *not assessable*. The highest evidence level, an established health effect, is given if the effect has been replicated in independent studies and if causation through RF EMF exposure is biologically plausible. The Swiss approach focuses on the evaluation of health effects from RF EMF exposure exclusively, and considers only experimental studies in humans and epidemiological studies. In addition to the level of evidence the Swiss approach includes two more aspects, i.e. *relevance to health* and *exposure*, which allow a further differentiation in the evaluation of health effects.

Norbert Leitgeb presents another procedure that uses a categorical system for characterizing evidence: the approach of the German Commission on Radiation Protection (SSK) for the evaluation of potential health risks from ELF and RF EMFs. The highest level of evidence is described as *scientific evidence*, which requires that a relationship between EMF exposure and a health relevant outcome has been shown in reproduced studies of independent research groups and that a causal relationship is plausible by the overall scientific knowledge. The next evidence level is *justified scientific suspicion*, followed by *scientific indication*. The weakest evidence level is *inadequate data*, which includes contradictory outcomes of single studies or results from studies with questionable methodological quality. An import difference to the IARC approach is that the SSK does not confine itself to carcinogenicity, but also includes other endpoints in its evaluation, such as cognitive functions, sleep or blood parameters.

A different approach has been taken by the California EMF Program which addressed the potential health effects from power frequency magnetic fields. As Raymond Neutra explains in his chapter, the California EMF Program explicitly wanted to go beyond the IARC classification type of evaluation. It aimed at providing a quantitative characterization of the degree of certainty one would be willing to express regarding a causal relationship between ELF magnetic field exposure and effects on selected endpoints. This was accomplished by developing a *degree of certainty* scale ranging from 0 to 100 with numerically defined categories (e.g. "prone to believe", "strongly believe"). The methodological basis for that procedure was a qualitative Bayesian approach which focuses on how much more (or less) likely a specific pattern of experimental or epidemiological evidence would be under the causal hypothesis compared to the noncausal hypothesis.

The evidence maps approach, presented by Holger Schütz, Peter Wiedemann and Albena Spangenberg, differs from all approaches described above as it does not use a classification of evidence levels nor does it provide a quantitative expression of the degree of certainty for a causal relationship between exposure to a (potentially) hazardous substance or condition and the endpoints that are considered. Instead, evidence maps are designed to depict the reasons which lead experts to their conclusions when summarizing and evaluating the scientific evidence about a (potential) hazard. They provide a graphical representation of the arguments that speak for or against the existence of a causal relationship between exposure to a (potentially) hazardous substance or condition and the endpoints that are considered, as well as the conclusions that are drawn and the remaining uncertainties. The authors illustrate this approach using the example of cancer epidemiology for RF EMFs.

Parts II and III show how evidence is characterized in different research areas and describes procedures that are used to summarize and integrate evidence into an overall risk characterization. In this context, dealing with conflicting data is an integral (though challenging) task. But how do lay people make sense of conflicting data, how can uncertainties be communicated to the public? Are there any ethical guidelines that can help in making risk related decisions under uncertainty? Part IV (*Psychological and Ethical Aspects in Dealing with Conflicting Data and Uncertainty*) addresses these aspects.

Peter Wiedemann, Holger Schütz and Andrea Thalmann summarize what is known from psychological research on the perception of uncertainty and the communication about unclear risks. Their chapter starts with a brief discussion of uncertainty in risk assessment that provides the basis for identifying the critical issues for communicating uncertainty and understanding lay perception of uncertainty information. Topics addressed here are how lay people understand basic toxicological and epidemiological concepts, such as dose–response relationship or relative risk, how lay people integrate uncertainty information in their risk appraisals, and how different types of uncertainty descriptions (qualitative, quantitative and visual) affect their understanding of risk information. Another important factor in risk perception and risk communication is the role of context. Just as experts use their professional knowledge to evaluate evidence, lay people also rely on knowledge when evaluating evidence regarding a potential health risk. However, unlike the content-related expert risk knowledge, risk knowledge of lay people is, to a large extent, context-related. Thus, contextual effects, such as prior risk-related beliefs or the perceived trustworthiness of an information source, will affect how information about a hazard or risk is integrated into lay persons' risk appraisal.

Armin Grunwald examines the question whether ethics can provide guidance for dealing with unclear risk. Typical problems which emerge in situations of unclear risk include the acceptability and comparability of risks, the advisability of weighing up risks against opportunities, and the rationality of action under uncertainty. Grunwald examines five general ethical approaches which can and have been used to deal with such problems: the consequentialist approach, the principle of pragmatic consistency, the ethics of responsibility, the "projected time" approach and deontological advice. He concludes that these approaches are all limited in the guidance they can provide for dealing with unclear risks; however, they help structuring the problem space with regard to the normative issues involved. Grunwald then discusses two operative approaches which are often invoked as guiding principles in situations of unclear risk: the "precautionary principle" and the "principle of prudent avoidance", which both make (explicitly or implicitly) reference to ethical positions. Grunwald notes that both approaches transform the problem of normative guidance for dealing with unclear risks into the problem of establishing procedures for evaluating the available evidence concerning the risk.

In the final chapter of the book, Peter Wiedemann, Franziska Börner and Holger Schütz provide their view of what can be learned from the contributions in the book for communicating conflicting evidence for risk characterization. Their recommendations are structured by six guiding principles that are especially relevant for communicating in scientific evidence in risk assessment: prudence, transparency, impartiality, reasonableness, clarity and responsibility.

References

1 National Research Council (1983) *Risk Assessment in the Federal Government: Managing the Process*, National Academy Press, Washington, DC.
2 International Programme On Chemical Safety (2004) *Risk Assessment Terminology – Part 1 and Part 2*, World Health Organization, Geneva.
3 EPA (2000) *Risk Characterization Handbook (EPA 100-B-00-002)*, US Environmental Protection Agency, Washington, DC.
4 IARC (2006) *Preamble to the IARC Monographs. IARC Monographs Programme on the Evaluation of Carcinogenic Risks to Humans*, International Agency for Research on Cancer, Lyon [http://monographs.iarc.fr/ENG/Preamble/CurrentPreamble.pdf] [Retrieved: 24.09.2007].
5 Frewer, L.J., Raats, M. M. and Shepherd, R. (1993) Modelling the media: the transmission of risk information in the British quality press, *IMA Journal of Mathematics Applied in Business & Industry*, **5**, 235–247.
6 Schulz, W. (1976) *Die Konstruktion von Realität in den Nachrichtenmedien: Analyse der aktuellen Berichterstattung [The Construction of Reality in the News Media. An Analysis of the News Coverage]*, Alber, Freiburg.
7 Dearing, J.W. (1995) Newspaper coverage of maverick science: Creating controversy through balancing, *Public Understanding of Science*, **4**, 341–361.
8 Tickner, J.A. (2003) Introduction. In *Precaution: Environmental Science, and Preventive Public Policy*, (ed. J. A. Tickner), pp xiii–xix, Island Press, Washington, DC.

9 Foster, K.R., Vecchia, P. and Repacholi, M.H. (2000) Science and the precautionary principle, *Science*, **288**, 979–981.
10 Goklany, I.M. (2002) From precautionary principle to risk–risk analysis, *Nature Biotechnology*, **20**, 1075.
11 Marchant, G.E. (2003) From general policy to legal rule: aspirations and limitations of the precautionary principle, *Environmental Health Perspectives*, **111**, 1799–803.
12 Wertheimer, N. and Leeper, E. (1979) Electrical wiring configurations and childhood cancer, *American Journal of Epidemiology*, **109**, 273–84.
13 NIEHS (1999) *Health Effects from Exposure to Power-Line Frequency Electric and Magnetic Fields (NIH Publication No. 99-4493)*, National Institute of Environmental Health Sciences, National Institutes of Health, Research Triangle Park, NC.
14 WHO (2007) *Environmental Health Criteria Monograph No. 238: Extremely Low Frequency (ELF) Fields*, World Health Organization, Geneva [http://www.who.int/peh-emf/publications/elf_ehc/en/index.html] [Retrieved: 16.02.2007].
15 IARC (2002) *IARC Monographs on the Evaluation of Carcinogenic Risks to Humans: Volume 80. Non-ionizing Radiation, Part 1: Static and Extremely Low-frequency (ELF) Electric and Magnetic Fields*, International Agency for Research on Cancer, Lyon.
16 Burgess, A. (2004) *Cellular Phones, Public Fears, and a Culture of Precaution*, Cambridge University Press, Cambridge.
17 ICNIRP (1998) Guidelines for limiting exposure to time-varying electric, magnetic, and electromagnetic fields (up to 300 GHz), *Health Physics* **74**, 494–522.
18 Röösli, M. and Rapp, R. (2003) *Hochfrequente Strahlung und Gesundheit*, Bundesamt für Umwelt, Wald und Landschaft (ed.), Umwelt-Materialien Nr. 162, Bern [http://www.bafu.admin.ch/php/modules/shop/files/pdf/php1Rr5fL.pdf] [Retrieved: 16.04.2007].
19 Health Council of The Netherlands (2002) *Mobile Telephones – An Evaluation of Health Effects*, Health Council of The Netherlands (Gezondheidsraad), The Hague.
20 IEGMP (2000) *Mobile Phones and Health*, Independent Expert Group on Mobile Phones. National Radiological Protection Board, Chilton [http://www.iegmp.org.uk/] [Retrieved: 04.05.2007].
21 RSC (1999) *A Review of the Potential Health Risks of Radiofrequency Fields from Wireless Telecommunication Devices (RSC/EPR 99-1)*, Royal Society of Canada, Ottawa.
22 Strahlenschutzkommission (2001) *Grenzwerte und Vorsorgemaßnahmen zum Schutz der Bevölkerung vor Elektromagnetischen Feldern – Empfehlung der Strahlenschutzkommission*, Strahlenschutzkommission, Bonn [http://www.ssk.de/werke/volltext/2001/ssk0103.pdf] [Retrieved: 27.09.2007].
23 Zmirou, D. (2001) Les Telephones Mobiles, Leurs Stations de Base et la Sante Etat des connaissances et recommandations, Rapport au Directeur Général de la Sante, Paris.

I
From Scientific Analysis to Risk Policy

2
Risk Assessment and Risk Communication for Electromagnetic Fields: A World Health Organization Perspective

T. Emilie van Deventer and Kenneth R. Foster

2.1
Introduction

Identifying, communicating and addressing health risks in the environment is a central concern of the World Health Organization (WHO), which aims to help Member States achieve safe, sustainable and health-enhancing human environments, protected from biological, chemical and physical hazards. The health impact from exposure to electromagnetic fields (EMFs) in the environment and workplace falls within WHO's mandate, where health is defined as a "state of complete physical, mental and social well being and not merely the absence of disease or infirmity" [1].

Assessing the possible health effects of electromagnetic fields at levels encountered in ordinary occupational and nonoccupational environments is challenging. The scientific literature related to possible health effects of nonionizing EMFs is large and frequently inconsistent. Tens of thousands of scientific papers have been published over a period of more than 50 years, which vary widely in frequency range covered, methodology, biological endpoint and relevance to human health. Some papers report health effects from exposure to EMFs; others find no effects. Often the effects are small, frequently at the edge of statistical significance, and the studies are highly variable in quality. This literature has helped contribute to a high level of concern by members of the public about possible risks from fields associated with modern technologies such as mobile phones, power distribution systems, visual display terminals and radar installations. All of these factors create obvious problems in risk analysis.

In this chapter we describe the process used by WHO for identifying and characterizing the evidence for hazards of static and extremely low-frequency (ELF) electric and magnetic fields, and the associated activities relating to risk communication. The risk assessment is discussed with reference to two recent Environmental Health Criteria (EHC) documents, one for static fields [2] (referred to as static-EHC) and one for ELF fields [3] (hereafter referred to as ELF-EHC).

The Role of Evidence in Risk Characterization: Making Sense of Conflicting Data.
Edited by Peter M. Wiedemann and Holger Schütz
Copyright © 2008 WILEY-VCH Verlag GmbH & Co. KGaA, Weinheim
ISBN: 978-3-527-32048-6

Static electric and magnetic fields are associated with some industrial and medical equipment, while ELF fields are commonly encountered in the environment from electrical power distribution systems and in close proximity to electrical appliances. We summarize WHO's outputs in terms of risk communication and place these efforts in a context with respect to three different models for risk communication.

2.2
Conceptual Framework for Risk Assessment

WHO employs a conceptual framework for assessing health risks to humans that closely follows that of the famous "Red Book" published by the US National Research Council in 1983 [4], which was developed in an attempt to separate, however imperfectly, "science" from "policy" aspects of analysis and management of risk. In the Red Book approach, which has been influential in risk analysis around the world, risk analysis is separated into three components: risk assessment (evaluating evidence regarding risks), risk management (controlling risks) and risk communication. Appendix 1 defines relevant terms from WHO sources.

In 1973, WHO began its EHC Programme with several objectives, among which are: "to assess information on the relationship between exposure to environmental pollutants and human health, and to provide guidelines for setting exposure limits" (Preamble [3]). A major output of this program is the series of EHC monographs which are critical reviews of the international scientific literature on the effects of chemicals or other potentially hazardous biological and physical agents on human health and the environment. The formal development process and format of these documents is closely specified by WHO policies.

2.3
EHC on EMFs

WHO's assessment of health risks produced by radiation-emitting technologies falls within the responsibilities of the Radiation and Environmental Health Unit (RAD). The International EMF Project, managed by RAD, was formed in 1996 as a multinational, multidisciplinary effort to assess health and environmental effects of exposure to nonionizing electromagnetic fields and radiation (0–300 GHz), to create and disseminate information appropriate to human health risk assessment for EMFs, and to provide technical assistance in strengthening national capacities for the sound management of EMF health risks.

It has been more than a decade since a series of EHC documents on EMFs was released by WHO [EHC 16 Radiofrequency and microwaves (1981); EHC 35 ELF fields (1984); EHC 69 Magnetic fields (1987); EHC 137 EMFs (300 Hz to 300 GHz) (1993)]. The EMF Project is now approximately two-thirds of the way through the development of updated and revised EHC documents on EMFs. Already published are volumes on static fields [2] and ELF fields [3]; the final volume on radiofrequency fields is anticipated to be released in 2010.

The Preambles of the static- and ELF-EHCs describe the common goals of the monographs:

> The EHC monographs are intended to assist national and international authorities in making risk assessments and subsequent risk management decisions. They represent an evaluation of risks as far as the data will allow and are not, in any sense, recommendations for regulation or standard setting. These latter are the exclusive purview of national and regional governments. However, the EHCs do provide bodies such as [the International Commission on Non-Ionizing Radiation Protection] with the scientific basis for reviewing their international exposure guidelines.

The phrase "as far as the data will allow" implies a scientifically conservative approach. The intended audience consists of health specialists and scientific experts in Member States and policy makers who are ultimately deciding on risk management issues at the national level.

The process includes the appointment of a Task Group, which is charged with assessing risks to health from exposure to ELF electric and magnetic fields in the given frequency range (static fields, 0 Hz; ELF, >0–100 kHz; this should not be confused with the ELF frequency range of 3–30 Hz as typically used in the engineering literature). The Task Group is charged with reaching agreement by consensus, with the understanding that its final conclusions and recommendations cannot be altered after the final Task Group meeting. The members of the Task Group are experts in EMF research who represent themselves as individual scientists and not their organizations, varied in expertise and views, gender and geographical distribution. Members of the Task Group are approved by the Assistant-Director General of WHO and all the authors, consultants or advisers participating in the preparation of the EHC monographs sign a conflict of interest statement. Observers from industry, nongovernmental organizations, and relevant national and international agencies may be invited to attend the meetings, but can only speak at the invitation of the Chairperson and do not participate in the final evaluation. Additional input is provided by as many as 150 individuals around the world who are sent drafts of the EHC monographs to review.

The EHC documents are intended as a critical review of the literature, but not a summary of every paper in their fields, and they encompass all health outcomes. In addition to primary scientific literature, the authors of the EMF monographs drew on previous reviews by expert bodies, including a previous monograph on ELF fields by the International Agency for Research on Cancer (IARC, established under the auspices of WHO) and expert assessments by government agencies.

In arriving at their conclusions, the members of the Task Groups used a weight-of-evidence approach. The Preamble of the ELF-EHC describes this approach:

> All studies, with either positive or negative effects, need to be evaluated and judged on their own merit, and then all together in a weight of evidence approach. It is important to determine how much a set of evidence changes the probability that exposure causes an outcome. Generally, studies must be

replicated or be in agreement with similar studies. The evidence for an effect is further strengthened if the results from different types of studies (epidemiology and laboratory) point to the same conclusion.

2.3.1
ELF Fields

The 450-age ELF-EHC is organized into chapters that consider specific disease categories encompassing neurobehavior, neuroendocrine system, neurodegenerative disorders, cardiovascular disorders, immunology and hematology, reproduction and development, and cancer. Other chapters discuss dosimetry, biophysical mechanisms and other topics.

The opening chapter summarizes the major conclusions, Chapter 12 provides a comprehensive health risk assessment, while Chapter 13 describes protective measures. Probably the most closely read chapters will be those discussing possible links between exposure to ELF magnetic fields and illness, particularly between exposure to power frequency fields (50 or 60 Hz) and childhood leukemia. The possible link between ELF fields and childhood cancer was first reported in 1979 by Wertheimer and Leeper [5], and has been the subject of major research efforts in the subsequent years, including many epidemiology and animal studies. In 2002, IARC classified ELF magnetic fields as "possibly carcinogenic to humans", based on limited evidence of carcinogenicity from epidemiology studies in humans and less than sufficient evidence for carcinogenicity in experimental animal studies. Five years later, the ELF-EHC reaffirmed this classification, but also noted the weaknesses in the epidemiologic evidence to date and the continued lack of supporting animal data. "Thus, on balance", it concluded, "the evidence related to childhood leukemia is not strong enough to be considered causal".

Major conclusions regarding acute and chronic biological effects include (with quotes from the document itself):

Acute effects
Acute biological effects have been established for exposure to ELF electric and magnetic fields in the frequency range up to 100 kHz that may have adverse consequences on health. Therefore, exposure limits are needed. International guidelines exist that have addressed this issue. Compliance with these guidelines provides adequate protection for acute effects.

Chronic effects
Scientific evidence suggesting that everyday, chronic low-intensity (above 0.3–0.4 µT) power-frequency magnetic field exposure poses a health risk is based on epidemiological studies demonstrating a consistent pattern of increased risk for childhood leukaemia. Uncertainties in the hazard assessment include the role that control selection bias and exposure misclassification might have on the observed relationship between magnetic fields and childhood leukaemia. In addition, virtually all of the laboratory evidence and the mechanistic evidence fail to support a relationship between low-level ELF magnetic fields and changes in biological function or disease

status. Thus, on balance, the evidence is not strong enough to be considered causal, but sufficiently strong to remain a concern . . . .

The ELF-EHC regarded evidence for other links between exposure to ELF magnetic fields and health effects (including other childhood and adult cancers, suicide, reproductive, and development disorders) as "much weaker than for childhood leukemia":

> A number of other diseases have been investigated for possible association with ELF magnetic field exposure. These include cancers in both children and adults, depression, suicide, reproductive dysfunction, developmental disorders, immunological modifications and neurological disease. The scientific evidence supporting a linkage between ELF magnetic fields and any of these diseases is much weaker than for childhood leukaemia and in some cases (for example, for cardiovascular disease or breast cancer) the evidence is sufficient to give confidence that magnetic fields do not cause the disease.

In short, the ELF-EHC concluded that presently available scientific evidence did not establish the presence of hazards from ELF electric or magnetic fields at levels encountered in ordinary occupational or nonoccupational settings. This review is, undoubtedly, the most painstaking and thorough risk assessment currently available regarding possible health effects of ELF electric and magnetic fields.

2.3.2
Static Fields

The 369-page static-EHC in its summary identified hazards from static electric and magnetic fields under very high exposures. For electric fields, it is noted that:

> There are no studies . . . from which any conclusions on chronic or delayed effects can be made . . . On the whole, the results suggest that the only adverse health effects are associated with direct perception of fields and discomfort from microshocks.

Most of the available data relates to magnetic field exposure, where it is concluded that:

> . . . the available evidence from epidemiological and laboratory studies is not sufficient to draw any conclusions with regard to chronic and delayed effects. . . Their carcinogenicity to humans is therefore not at present classifiable.

A few cardiovascular effects were noted at field levels above 8 T (very strong fields such as might be produced by high-field magnetic resonance imaging scanners) and the EHC noted that biologically significant electric fields could be induced by movement of the body in a strong static magnetic field. However, the review found that the evidence for most reported effects the review found to be insufficient to "draw firm conclusions" about possible health risks.

2.4
Comparison Between the WHO ELF-EHC and the California Report

It is interesting to compare the ELF-EHC report with an earlier (2002) review by the California Department of Health Services (DHS) (Ref. [6], see also Chapter 12 of this volume), which covered the scientific literature up to about 2000. The report was written by three DHS employees, all senior scientists with expertise in epidemiology.

While the ELF-EHC document was a conventional literature review, the California review took an approach that was reminiscent (but not equivalent to) a Bayesian analysis [7] [A Bayesian analysis would require separate consideration of two probabilities: the probability of an observation or experimental result if the hypothesis being examined (e.g. exposure causes disease) is true and second one if it is false. That would require a far more extensive examination of the data than was presented in the California study and the probabilities themselves are most likely not available in the literature.] In this process, each of the three reviewers initially stated their "priors" or initial degrees of confidence that various reported associations between EMF exposure and disease were "real", i.e. causal. Then, for each health endpoint, the three DHS scientists considered various arguments for or against causality. At the end of the process, each reviewer stated a numerical score indicating his or her final level of confidence that a hypothesized association between exposure and disease was causal, i.e. that the exposure caused the disease. The reviewers also described their conclusions in narrative terms, e.g. describing their confidence as being "close to the dividing line between believing and not believing" or "prone not to believe" that EMF exposure caused various health problems.

In sharp contrast to the ELF-EHC, which considered a range of evidence including mechanistic, *in vitro* and *in vivo* studies, the California report focused primarily on epidemiologic evidence, and placed relatively little weight on animal and mechanistic studies. "Overall", the California report concluded, "the prior of the review team was little changed by biophysical arguments".

By comparison, the ELF-EHC review included an extensive chapter that reviewed biophysical and mechanistic studies. The "absence of an identified plausible mechanism", the ELF-EHC concluded, "does not rule out the possibility of adverse health effects, but it does increase the need for stronger evidence from biology and epidemiology". This lack of a plausible mechanism for bioeffects of ELF fields, together with the lack of supporting animal evidence, have been major stumbling blocks for many scientists in accepting that exposure to low-level ELF fields could cause health effects in humans.

Not surprisingly given these differences, the two reviews differ strikingly in their conclusions. The California review concluded:

> To one degree or another, all three of the DHS scientists are inclined to believe that EMFs can cause some degree of increased risk of childhood leukemia, adult brain cancer, Lou Gehrig's Disease, and miscarriage ... For adult leukemia, two of the scientists were "close to the dividing line between believing or not believing" and one was "prone to believe" that EMFs cause some degree of increased risk.

This contrasts with the "inadequate" evidence that the ELF-EHC reported for a link between ELF magnetic fields and childhood leukemia, and the "still weaker" evidence for other health effects.

Owing to its novel approach, the California report is difficult to compare with other health risk assessments on EMF. The IARC employs specified weight-of-evidence criteria in its carcinogen risk assessment and the IARC classification of ELF magnetic fields as a "possible human carcinogen" has a clear meaning in terms of experts' evaluation of scientific evidence (i.e. limited epidemiologic and lack of supporting animal data). In the California study the relationship is more complex. However, it makes explicit the large element of human judgment that is involved in risk analysis and a comparison of these two assessments illustrates the extent that judgments can vary among otherwise well-qualified individuals.

2.5
Communicating about Risks of EMFs

WHO in general, and the EMF Project in particular, principally supports the needs of the health ministries of Member States and one of its aims is to help the relevant national authorities develop their risk communication strategies. The Project communicates about risk through several mechanisms.

One mechanism is through Fact Sheets, which are short papers (3–4 pages) that discuss current EMF topics in lay-oriented language, available both in printed form and electronically through the EMF Project website (http://www.who.int/docstore/peh-emf/publications/facts_press/fact_english.htm). Simultaneously with the publication of the EHCs, the EMF Project issued Fact Sheets on static and ELF fields that summarized the conclusions of the longer scientific documents.

A more general approach to risk communication and EMF was taken in 2002 when the EMF Project published a 66-page booklet on risk communication *Establishing a Dialogue on Risks from Electromagnetic Fields* [8], which is available both as a glossy brochure and electronically through the EMF Project website (http://www.who.int/peh-emf/publications/risk_hand/en/index.html). The three chapters review the present evidence related to EMFs and public health, basic principles of risk communication, and a brief discussion of science-based and precautionary approaches to regulation of public exposure to electromagnetic fields. This booklet has been of obvious interest to a number of countries, as evidenced by its translation into several languages and broad dissemination (e.g. the Italian government translated the booklet and distributed 50 000 copies to local government officials.)

The approach of the manual is conveyed by some opening lines:

> Many governmental and private organizations have learned a fundamental, albeit sometimes painful lesson; that it is dangerous to assume that impacted communities do not want, or are incapable of meaningful input to decisions about siting new EMF facilities or approving new technologies prior to their use. It is therefore crucial to establish a dialogue between all individuals and groups impacted by such issues. The ingredients for effective dialogue

include consultation with stakeholders, acknowledgement of scientific uncertainty, consideration of alternatives, and a fair and transparent decision-making process. Failure to do these things can result in loss of trust and flawed decision-making as well as project delays and increased costs . . . .

Risk communication is therefore not only a presentation of the scientific calculation of risk, but also a forum for discussion on broader issues of ethical and moral concern. Environmental issues that involve uncertainty as to health risks require supportable decisions. To that end, scientists must communicate scientific evidence clearly; government agencies must inform people about safety regulations and policy measures; and concerned citizens must decide to what extent they are willing to accept such risk. In this process, it is important that communication between these stakeholders be done clearly and effectively.

Other organizations have used the scientific information in their own efforts at risk communication. For example, GreenFacts, a nongovernmental organization, developed an attractive website for the nonspecialists that poses the question "What is known so far about potential health consequences? [of static fields]?" (http://www.greenfacts.org/en/static-fields/index.htm) based on the content of the Static-EHC monograph.

2.6
Discussion

It is interesting to consider this WHO approach to risk assessment and communication (in the present context, the EHCs, Fact Sheets and other work by the EMF Project related to risk communication) in a broader context of risk communication as studied by social scientists. Van Zwanenberg and Millstone summarized and criticized three models of risk communication with respect to a different issue – government communications with the public about risks of bovine spongiform encephalopathy [9]. While the models pertain to risk communication, according to these authors, they are "tied to a broader set of consistent ideas and assumptions through which analysts and practitioners have understood the nature of science-based policy-making".

In the *technocratic approach*, scientists are (all quotes are from the van Zwanenberg and Millstone paper)

> . . . presumed to be in possession of "the truth", or at least reliable knowledge about risk (defined as an objective probability of harm), whereas the general public was understood or represented as being at best ignorant, and at worst possessed of false and unscientific beliefs. From that perspective, risk communication was seen as the attempt to provide science-based representations of risk that were sufficiently simplified to be readily transferable into the minds of the general public in order to diminish their ignorance or to displace alternative representations of risk.

> [Risk communication was] conceived . . . as a tertiary consideration (i.e. subsequent to, first, the assessment of risk and, second, the identification of

policy responses). Risk communication was thus a one-way, top-down process running from the experts to the government and thence to industrial stakeholders and the general public . . . . [R]epresentations of technical risk, provided by the officially selected scientific experts, were deemed to be unproblematically correct and adequate.

In the *decisionist* approach:

> . . . risk communication is understood as a two-way rather than a one-way process. On the one hand, technical attributes of risk are communicated in one direction, from the experts to the government and thence to the general public. On the other hand, the media and other mechanisms have a legitimate role to play in helping policy-makers to understand the conflicting concerns and interests of different social groups and their varying willingness to tolerate different kinds of risks in exchange for different kinds of benefits. From the point of view of policy-makers, the public need to be persuaded that risk management decisions are prudent and fair, but that can only be accomplished if policy-makers understand how the public view issues such as prudence and fairness as they apply to the issues at stake. Contemporary official risk communication guidelines typically stress the importance of both understanding the attitudes to risk of affected and interested citizens and of incorporating those views and preferences into policy.

> . . . Policy-makers who acknowledge that risk communication, under the decisionist approach, is a two-way process typically conceive of the relationship between science and policy as one in which science is necessary but not sufficient for policy decision-making.

The *deliberative* approach, said to be the most recent of the three models, arose from the recognition of social scientists that the process of risk assessment necessarily is coupled to social values; thus in this view the goal of the Red Book to separate "science" from "values" cannot be accomplished. Advocates of this approach urge that nonscientist stakeholders be involved in the risk assessment process itself.

> [T]here are often scientific uncertainties involved in assessing risk issues, and . . . scientists interpret shared bodies of evidence in differing and conflicting ways. . . . A key feature of a deliberative approach is that risk communication . . . needs to involve dialogue about the definition and analysis, as well as the evaluation, of any particular risk issue.

As an example of this approach, the (US) National Research Council [10] suggests that:

> . . . deliberative processes – involving all interested and affected parties – are necessary when deciding which types of harm to analyze, deciding how to describe scientific uncertainty and disagreement, analyzing evidence, generating policy options, and deciding on policy outcomes.

This WHO approach to risk analysis might be represented by some commentators as a paradigmatic example of a technocratic approach to managing and communicating risk. However, the criticisms raised by van Zwanenberg and Millstone [9] in their discussion of the "technocratic model" of risk assessment hardly apply to the risk communication of WHO. As an international health agency, WHO's mandate is to provide reliable assessments of scientific and medical evidence to Member States. The diverse social and cultural values and situations, which vary tremendously in different Member States, cannot be easily included in risk analysis at the stage at which WHO operates. However, cultural and social aspects of risk can (and should) be dealt with on a regional or national level, along with other economic and development aspects. An example is the series of "deliberative consensus conferences on science and technology policy" that are held in a number of countries (http://www.loka.org/pages/worldpanels.htm). This approach is not excluded by WHO approach to risk analysis and communication, but can be complementary to it.

Moreover, there is no claim of infallible knowledge in the EHCs, which is another of van Zwanenberg and Millstone's [9] criticism of technocratic approaches to risk communication. All of the EHC documents of WHO present themselves as considered assessment of scientific evidence reflecting a consensus of expert views – and not as unambiguous "truth" or facts that are "unproblematically correct and adequate."

Indeed, in all EMF Project documents, there is an emphasis on scientific uncertainty and gaps in scientific knowledge. Thus the ELF-EHC recommended "Government and industry should promote research programs to reduce the uncertainty of the scientific evidence on the health effects of ELF field exposure". The EHCs of the EMF Project have included sets of research recommendations, which have been implemented and resulted in funded research in several Member States.

In closing, characterizing evidence in EMF risk assessment is a complex endeavor due to the complexity of the information at hand. With the rapid deployment of new technologies employing electromagnetic fields, it is a challenge to identify potential health risks, particularly those that might emerge after long use of the technologies, and to communicate with the public and governments what is known and levels of scientific uncertainty regarding any potential health hazard. However, there is a similar need to identify and communicate on the health benefits of these technologies so as to balance those risks and benefits and determine their overall health impact.

Appendix 1

Definitions of risk analysis as used in this chapter (descriptions of selected key generic terms used in chemical hazard/risk assessment; International Programme on Chemical Safety Joint Project with OECD on the Harmonization of Hazard/Risk Assessment Terminology: http://www.who.int/ipcs/publications/methods/harmonization/definitions_terms/en/):

- *Risk assessment.* A process intended to calculate or estimate the risk to a given target organism, system or (sub)population, including the identification of attendant uncertainties, following exposure to a particular agent, taking into account the inherent characteristics of the agent of concern as well as the characteristics of the specific target system.

- *Risk management.* The decision-making process involving considerations of political, social, economic and technical factors with relevant risk assessment information relating to a hazard so as to develop, analyze and compare regulatory and nonregulatory options, and to select and implement appropriate regulatory response to that hazard. Risk management comprises three elements: risk evaluation, emission and exposure control, and risk monitoring.
- *Risk communication.* The interactive exchange of information about (health or environmental) risks among risk assessors, managers, news media, interested groups and the general public.

The risk assessment process includes four steps:

- *Hazard identification* is the identification of the type and nature of adverse effects that an agent has as inherent capacity to cause in an organism, system or (sub)population.
- *Hazard characterization* (related term: dose–response assessment) is the qualitative and, wherever possible, quantitative description of the inherent properties of an agent or situation having the potential to cause adverse effects. This should, where possible, include a dose–response assessment and its attendant uncertainties.
- *Exposure assessment* is the evaluation of the exposure of an organism, system or (sub)population to an agent (and its derivatives).
- *Risk characterization* is the qualitative and, wherever possible, quantitative determination, including attendant uncertainties, of the probability of occurrence of known and potential adverse effects of an agent in a given organism, system or (sub)population under defined exposure conditions.

References

1 WHO (1948) *Preamble to the Constitution of the World Health Organization as adopted by the International Health Conference, New York, 19–22 June, 1946; signed on 22 July 1946 by the representatives of 61 States (Official Records of the World Health Organization, no. 2, p. 100) and entered into force on 7 April 1948*, World Health Organization, Geneva [http://www.searo.who.int/LinkFiles/About_SEARO_const.pdf] [Retrieved: 16.02.2007].

2 WHO (2006) *Environmental Health Criteria Monograph No. 232: Static Fields*, World Health Organization, Geneva [http://www.who.int/peh-emf/publications/reports/ehcstatic/en/index.html] [Retrieved: 16.02.2007].

3 WHO (2007) *Environmental Health Criteria Monograph No. 238: Extremely Low Frequency (ELF) Fields*, World Health Organization, Geneva [http://www.who.int/peh-emf/publications/elf_ehc/en/index.html] [Retrieved: 16.02.2007].

4 National Research Council (1983) *Risk Assessment in the Federal Government: Managing the Process*, National Academy Press, Washington, DC.

5 Wertheimer, N. and Leeper, E. (1979) Electrical wiring configurations and

childhood cancer, *American Journal of Epidemiology*, **109**, 273–284.

6 Neutra, R., DelPizzo, V. and Lee, G.M. (2002) *An Evaluation of the Possible Risks from Electric and Magnetic Fields (EMFs) from Power Lines, Internal Wiring, Electrical Occupations and Appliances*, California Department of Health Services, Oakland, CA [http://www.dhs.ca.gov/ehib/emf/RiskEvaluation/riskeval.html] [Retrieved: 16.02.2007].

7 Howson, C. and Urbach, P. (1993) *Scientific Reasoning: The Baysian approach*, Open Court, Chicago, IL.

8 WHO (2002) *Establishing a Dialogue on Risks from Electromagnetic Fields*, World Health Organization, Geneva.

9 van Zwanenberg, P. and Millstone, E. (2006) Risk communication strategies in public policy-making. In *Health hazards and Public Debate: Lessons for Risk Communication from the BSE/CJD Saga*, (ed. C. Dora), World Health Organization, Geneva.

10 National Research Council (1996) *Understanding Risk. Informing Decisions in a Democratic Society*, National Academy Press, Washington, DC.

3
Characterizing Evidence and Policy Making

Evi Vogel and Ginevra Delfini

3.1
Introduction

Technological progress offers a whole range of benefits for society and allows economic development; at the same time, though, it is associated with risks. Acceptance or nonacceptance of risks within society can be triggered in many different ways. Policy based purely on scientific evidence is therefore not adequate to fully address societal needs. Evidence additional to the pure scientific facts, we shall call it "society-based evidence", is present which influences the process of policy making.

In this chapter we will take a look at how both science-based evidence and society-based evidence lead to policies, using the example of nonionizing radiation.

3.2
Science-based Evidence

Each country sets its own national standards regarding exposure, and therefore from time to time asks national expert committees to review the research data available and make recommendations if the existing limit values may be kept or have to be reconsidered.

Figure 3.1 shows how evidence from different sources, scientific as well as social, eventually leads to policies. The "policy drafters" act as a filter for all the different strands of information and influences. Usually "policy drafters" work in governmental agencies, federal offices or ministries. Their task is to prepare the papers which form the basis for the political process leading in the end to regulations, recommendations and other policies. To prepare policies, they mostly use recommendations of scientific expert committees, but also take account of other scientific sources and society-based evidence, trying to select the "signals" from the "noise".

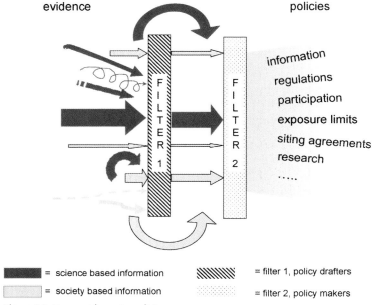

Figure 3.1 From evidence to policies.

Their filtered information as well as unfiltered information influence the politicians (the "policy makers") during the policy making process. The process itself again acts as another filter and its output are different kinds of policies.

In the case of nonionizing radiation [mostly called electromagnetic fields (EMFs)], the recommendations of the international expert group, the International Commission on Non-Ionizing Radiation Protection (ICNIRP), usually form the background for all these national committees and are used as a reference point. [The ICNIRP is a nongovernmental organization which is formally recognized by the World Health Organization (WHO) as the committee responsible for EMF guidelines. It evaluates all peer-reviewed scientific results available and gives recommendation on limit values which are reviewed periodically and updated as necessary.] National committees often also do compare their evaluations to the recommendations of other national committees. As a result of both the common basis (ICNIRP) and the interaction between countries it can be stated that the differences between the outcome of the national committees are usually small.

3.2.1
From Basic Research to Recommendations

Selecting the relevant scientific information and dealing with scientific uncertainties is the task of the committees responsible for reviewing literature and giving recommendations based on the existing scientific results and of expert judgment.

These committees provide policy drafters and policy makers with essential information which forms the "backbone" of any policy.

This task is difficult, mainly, due to two reasons: (i) the different scientific quality of publications and (ii) uncertainties.

To solve the first difficulty an answer must be given to questions like:

- Which are the relevant papers?
- Are the data reproducible?
- Is the interpretation of the authors correct?
- Are there methodological flaws, inconsistencies, etc.?

In principle this can be solved by considering only papers published in peer reviewed journals and which meet certain standards, e.g. standards of Good Laboratory Practice [1] and disregarding the others.

Concerning the second difficulty, uncertainty, it must be stressed that scientific uncertainties will always exist and that absolute certainty on matter of health risks cannot be achieved (see Box 3.1). In fact it is, in principle, impossible to demonstrate that a certain risk does not exist. This particularly represents a difficulty when communicating about the achieved results and therefore will always leave a certain uncertainty margin. It is important, however, that all known uncertainties are presented, discussed and explained as far as possible.

Box 3.1
A Categorization of Uncertainties (Adapted from Ref. [4])

Uncertainties can be put into different categories:

- Uncertainties that science is aware of and can deal with, e.g. related to the confidence interval of the results of an experiment – these uncertainties can be dealt with in a scientific manner.
- Uncertainties that science is aware of, but which cannot be measured, e.g. differing results in similar experiments – these uncertainties need a more comprehensive approach, in which expert knowledge and scientific experience play an important role.
- Uncertainties resulting from up to now unknown effects, reactions or developments. Different cultures deal differently with such imponderability and therefore it has to be coped with by the policy makers, outside the scientific framework.

A specific way of policy making in relation to uncertain scientific information deserves particular attention: the precautionary principle. The issue of when and how to use the precautionary principle, both within the European Union and internationally, has given rise to much debate. It is discussed thoroughly in Refs [2, 3] and we will not discuss it in detail here.

3.2.2
Officially Appointed Expert Committees and Self-appointed Experts

Clearly the value of the recommendations is also linked to the authority and the reliability of the committees which should be broadly acknowledged. Usually national expert committees are officially appointed and held responsible for their work. It is strongly advisable that different fields of scientific expertise are represented in such a group. The group should work according to a scientific method: assess working rules, and specify and explain the basis on which they formulate their recommendations. It should also be made clear if the scientific knowledge is divided, and on which points there is some disagreement and the reason for it. Also this information is part of the "science-based evidence".

Additional to the recommendations given by official expert committees, there often are recommendations given by single experts. Sometimes very renowned scientists, sometimes self-proclaimed experts give their personal opinion. It has to be kept in mind that these single opinions, although legitimate, tend to follow a selective choice, personal preferences and/or personal weighing of the facts, and therefore may contain more bias than committee opinions [5]. This information should not be all together discarded, but rather be given a proper weight in the process of policy making.

Interesting enough, the "scientific" authority and reliability mentioned above does not necessarily coincide with the "societal" authority and reliability: in fact, one of the first questions asked by the public is always: who chose the committee and what are the affiliations of its members? Only second comes the interest in the working rules of the group, although this, as discussed, is of prime importance from a scientific point of view. This different perception can lead to a societal preference for the opinion of a (single) "independent" expert, or expert group, even if less qualified on the specific scientific knowledge. If not acknowledged and not properly dealt with, this discrepancy can cause tension and result in non-acceptance of the resulting policies/decisions.

3.2.3
Communication of Recommendations

A challenge that should not be underestimated is the presentation and communication of recommendations based on scientific results. The committees must bear in mind that their recommendations will be interpreted in a political context and policy drafters as well as policy makers must respect that such recommendation might be provisional due to ongoing research. It is also important to remember that different groups in society will try to use parts of the recommendation to back up their own strategy.

Therefore, beside the necessity to clearly present the technical and scientific reasoning, the language used is very important. In different situations, and in front of different audiences, it could be necessary to use different wordings. This can be difficult and requires a thorough preparation, answering questions like:

- What does the audience already know about the subject?
- How can we elaborate on that?
- What kind of information does the audience expect or need?
- How far should scientific explanation go?
- How to avoid technically correct but unknown terminology?

Recommendations on risk communication especially concerning nonionizing radiation are given by WHO [3].

3.3
Society-based Evidence

Societal unrest is often a consequence of a not well defined, but persistent feeling of uncertainty in the presence of unknown technologies. Only relatively few people can cope with scientific uncertainties; most feel at least uneasy when they have to live with risks, even when the risks are small. The science of risk perception does reveal and explain the factors determining how we deal with different risks (see references in Ref. [3]), but nevertheless this knowledge will not help much to influence public acceptance of risks.

Also, trying to give scientific explanations usually is not enough: people want certainties, and what they get is "probabilities", "confidence intervals" and "error bars". For most of them this is not satisfying, but even more unnerving.

In the case of nonionizing radiation, societal unrest and fear of possible health effects is present in quite a few countries, even though the scientific basis for it is very weak or absent.

Apart from the general unrest of society that can be measured, such as by opinion polls (e.g. the German Federal Institute for Radiation Safety is doing yearly polls on the perception of EMFs and the results are published on their internet site www.bfs.de in the section on EMFs), there often are relatively small groups uttering strong opinions, in the case of EMFs both in favor of mobile phones (e.g. industry) and against it (e.g. "electrosmog" activists). Even though they are not representative of the majority opinions expressed in the above-mentioned polls, they may gain a lot of weight, e.g. when activist groups are active in the district of a single politician and/or when their opinion is enhanced by the media. These usually well-organized activist groups are devoted to their "cause" and therefore use all possibilities to aim at it: this often leads to strong, even though biased, statements. After all, they have no need to stick to the scientific truth, they use emotions and appeal to the feelings of their audience. It is therefore difficult to argue against these statements with a sober, scientific sound, but relatively weak message.

It is also important to realize that not all the signals coming from society are relevant and have to be taken into account by policy drafters and policy makers. For instance, amplification of information by pressure groups and media, who keep citing each other, can generate a lot of "noise" without actual facts. It is important that the "signal-to-noise" ratio is considered in the process of defining society-based evidence.

3.4 Policy Making

We believe that the keywords for successful policies are "confidence" and "conviction". If a decision has to be taken in the face of uncertainty, which is always the case concerning environmental risks such as EMFs, at least confidence has to be inspired by the process by which the decision was reached and by the people who take responsibility for these decisions, i.e. the politicians involved. If people trust the responsible politicians and the process, they will also accept that decisions were taken in the best possible way. In order to gain confidence it is important that the process of policy making is open, clear and transparent. This means that, on the one hand, alarming information, even if scientifically not relevant, should not be neglected because that would be interpreted as concealing important data; a serious reaction and a critical explanation should be given. On the other hand, the reaction to the same information should be appropriate and if the information is not relevant, this should not lead to essential changes in the policy (e.g. lowering exposure limits).

3.4.1 Role of Policy Drafters in Policy Making

The policy drafters have to take into account the scientific facts as well as the society-based evidence. The interaction between policy drafters and society is very complex, and differs in different countries and for different policy fields. Societal unrest is a problem which policy drafters should address in a proper way: if it is neglected with the hint that there is no scientific basis, the anger grows; if one "gives in" to the scientifically unjustified fear (by e.g. suggesting to lower exposure limits), the decision making process is undermined and it will be extremely difficult to justify ones choices.

3.4.2 Role of Politicians in Policy Making

Taking responsibility is the task of politicians and governors. They have to make sure that the future regulations take into account all relevant information. The political process a policy draft has to follow before it will become an effective regulation does make sure that different opinions are heard: the council of ministers has to agree, parliament has to agree and the relevant stakeholders such as unions, syndicates or industry are heard, too.

However, this path can be affected by many considerations such as party philosophy, need to profile, upcoming political agenda, personal "pet subject", parliament pressure, attitude in neighboring countries, etc. Political work is therefore strongly influenced by personal traits and preferences.

All these aspects will interfere with the original draft of the policy, changing it in positive or negative ways as seen from the standpoint of the policy drafter. It is important, however, that the process of policy making keeps its reliability, and that the reasoning behind policy measures can be seen and explained [6].

3.4.3
Policy Making and the Media

The task of the media is to inform people; in particular, new developments and techniques shall be presented and opinions reported. Ideally, media deal with information in a balanced way, and try to be as correct as possible in presenting facts and to be impartial in reporting opinions. In practice things are different, mainly because of two factors: information is spread not only by professionals and media need to "sell" their information [7].

The need to "sell" information means that there is some sort of pressure on professional journalists to polarize opinions and stress the differences more than the agreements.

Concerning scientific data, it often happens that, on the one hand, the media take up some scientific papers which seem to be controversial and tend to stress "off-opinions". On the other hand, little or no relevance is given to information confirming previous scientific results and therefore increasing their weight. Such information, although important from the scientific point of view, does not have "news" value and will not get as much coverage as controversial information. In a similar way the well balanced and internationally agreed recommendations of scientific committees are usually less reported than controversial opinions of single experts.

Additionally, some scientists enjoy attention of media and politics, which makes them susceptible to being exploited, possibly creating unbalanced information coverage.

Concerning the policy-making process, controversies and pressure groups are often given more space than ordinary policy-making work. Whenever, for example, policy makers neglect the request of a specific pressure group, this fact is likely to receive more coverage than the arguments why the request was neglected.

The other very important factor influencing the information flow during the whole process of policy making is that at the present time, and especially via the internet, media are directly available to everybody to spread information and not only to professionals. It often happens that pressure groups set up very attractive and well managed websites by which they try to get adhesions to their ideas.

All of this results in a skewed information profile in the media, which only experts can recognize as such. It is difficult to restore the balance and sometimes it does not pay to spend much effort in doing it, but it is very important for policy drafters and policy makers to be aware of it and to utilize the opportunities to clarify and complete the information.

3.4.4
Policies

In different countries large communication programs were started in order to make the process of policy making more transparent. Research on regulatory decision making helps to improve the process as such [8].

In the case of EMF the policies adopted beside regulations and limit values are manifold: information campaigns, research, measurement campaigns or participation programmes. In some countries the proposal of research projects, the contracting and the results, the recommendations of scientific committees, opinion polls and decisions of the council of ministers as well as the parliament protocols of hearings can be found on the relevant internet pages.

In setting up information campaigns and information programs on the scientific process as well as on the political process, care should be taken in order to address, beside the general public, the intermediate actors like teachers, physicians and community workers specifically. To provide them with correct information is very important and helps effectively to reduce societal unrest.

Different approaches concerning policies in the case of mobile telephony were presented and discussed during a workshop organized by the Joint Research Centre of the European Commission and the EMF-NET in May 2007 [9].

A measure which has proven quite successful in containing societal unrest concerning the siting of mobile phone antennas is the direct involvement of the public in the decision-making process about the site. Box 3.2 gives two examples for the involvement of the public in decision making (both examples were presented more extensively as case studies in Ref. [9]).

Box 3.2
Two Examples for the Involvement of the Public in Decision Making

(a) "Tenants Consent Procedure" (The Netherlands)

The "tenants consent procedure" was introduced in 2002, within the framework of a more comprehensive agreement between authorities and operators regarding siting procedures. When an operator wants to place a base station on the roof of an apartment building, not only the owner(s) of the building give permission, but also the tenants (and actual users) of the apartments are informed, consulted and asked their opinion. It is agreed that the placing of the antenna will not go on, if the majority (50% + 1) of the tenants is against it.

To ensure the impartial character of it, the full procedure, including the counting of the votes, is run by an independent agency. The "tenant consent procedure", gave positive results in addressing the local unrest and most sites have been accepted. On the average, only about 40% of the tenants actually did use the opportunity to vote. Although authorities increased their information efforts, the percentage of denied placing on the roof of apartment buildings did not change much in time and is seems stabilized at about 30%.

At the moment new measures, beside the "tenants consent procedure", are being studied in order to promote involvement and responsibility of the public in a different way.

(b) "Bavarian Mobile Phone Pact" (Germany)

The mobile phone providers in Bavaria, the Association of the Bavarian Communities, the Association of the Bavarian Counties and the State Government in 2002 signed the so-called "Mobilfunkpakt" ("Mobile Phone Pact"). It allows the communities to participate in the siting of mobile phone stations. They may suggest up to three options for a site which then have to be checked by the provider. In cities with more than 50 000 inhabitants this process is done in round table discussions, involving the main stakeholders and often also representatives of interest groups. In smaller communities the pact gives a timeline for a dialogue process between the community and the mobile phone providers.

The yearly reports on the progress of the "Mobilfunkpakt" show that the pact partners are content and that now less than 10% of the mobile phone base stations in Bavaria are built in dissent with the respective community.

3.5 Conclusions

The interactions between the different actors in society are complex to describe and to influence. Especially when dealing with uncertainties it is important that it is understood that different actors have different roles.

Scientists deliver scientific evidence, including all the uncertainties which are part of it. Expert committees review all the relevant studies, define frameworks for doing so, and explain the reasoning for their conclusions and recommendations. Policy drafters are responsible for understanding the information delivered by science and expert committees, and translating it into appropriate proposals for policies. To do so, they must, on the one hand, formulate strict regulations on the basis of scientific evidence, but also, on the other hand, carefully evaluate other signals, beside scientific evidence, coming both from science (e.g. uncertain data) and from society (e.g. unrest). From these signals a complementary kind of evidence should be extracted and accommodated within policy.

In the political process of policy making different stakeholders are heard and different politicians have a say. The decision making does involve a lot of very personal views and traits. The fact that so many individuals are involved in the political process provides by itself a guarantee that no-one's needs will be neglected. In the end, politicians and governors are responsible for decision making, and thus also for decisions about how to deal with uncertainties, risks and inadequacies, as well as how to deal with societal unrest. Their role also involves publicly announcing and defending the decision taken, in some countries new regulations have to stand a trial before a court.

Standards and limit values should adhere to the scientific recommendation, the "science-based evidence", because this is the only reasonable and trustworthy way to protect society against known health effects. On the other hand, policy making is broader than just formulating regulations on exposure limits – it has to take into account "society-based evidence". Especially when dealing with societal unrest different instruments have been proven successful: organizing public participation procedures, round tables, hearings, improving information and communication, offer the monitoring of EMFs, measurement campaigns and research programmes are all possible flanking measures which take account of special needs of society

In short, although science helps to find a scientific solution (exposure limit) to a concrete problem (health effect), societal demands have to be met too. Dealing with the fears and insecurities of a society, but also with its needs and requests, is part of the modern way of taking responsibility – this is what policy drafters as well as policy makers are asked to do.

References

1 Repacholi, M. and Van Deventer, E. (2003) WHO framework for developing EMF standards. In *Proceedings of the International Conference on Non-Ionizing Radiation at UNITEN (ICNIR 2003)*, 20–22 October, Kuala Lumpur [http://www.who.int/peh-emf/meetings/archive/en/paper17repacholi.pdf] [Retrieved: 13.04.2007].

2 CEC (2000) *Communication from the Commission on the Precautionary Principle. COM (2000).* Commission of the European Communities, Brussels [http://ec.europa.eu/ dgs/health_consumer/library/pub/pub07_en.pdf] [Retrieved: 13.04.2007].

3 WHO (2002) *Establishing a Dialogue on Risks from Electromagnetic Fields*, World Health Organization. Geneva.

4 Boeschen, S., Kastenhofer, K., Marschall, L., Rust, I., Soentgen, J. and Wehling, P. (2006) Scientific cultures of non-knowledge in the controversy over genetically modified organisms (GMO) – the cases of molecular biology and ecology *Gaia – Ecological Perspectives for Science and Society*, **15**, 294–301.

5 Schütz, H. and Wiedemann, P.M. (2005) How to deal with dissent among experts. Risk evaluation of EMF in a scientific dialogue, *Journal of Risk Research*, **8**, 531–545.

6 Weingart, P. (2006) Policy advice is an academic duty, *Humboldt Kosmos*, "*Science and Politics*" 87.

7 Larkin, J. (1999) Evaluating response actions, in *Proceedings of the International Seminar on EMF Risk Perception and Communication*, (eds M. Repacholi and A.M. Muc), pp. 137–150, World Health Organization, Geneva.

8 Presidential Commission on Risk Assessment and Risk Management (1997) *Framework for Environmental Health Management. Volumes 1 & 2,* Presidential Commission on Risk Assessment and Risk Management, Washington, DC [http://cfpub.epa.gov/ncea/cfm/pcrarm.cfm?ActType=default] [Retrieved: 13.04.2007].

9 Joint Research Centre, (2007) *Proceedings of the 2nd Workshop on EMF risk Communication:"Effective Risk Communication in the Context of Uncertainty*" 2–4 May, Stresa, Italy [http://www.jrc.ec.europa.eu/eis-emf/stresa2007.cfm] [Retrieved: 13.04.2007].

II
Making Sense of Conflicting Data: Evidence Characterization in Different Research Areas

4
Basic Principles and Evidence Characterization of the Data from Genetox Investigations

Günter Obe and Vijayalaxmi

4.1
Introduction

Genetic toxicology deals with the evaluation of adverse effects on cellular genetic material (DNA) following exposure to biological, chemical and physical agents. The effects of genotoxic agents have been investigated for several decades using different techniques and endpoints. Classical methods include the analysis of chromosomal aberrations (CAs), micronuclei (MN) and sister chromatid exchanges (SCEs) [1–3]. A simpler "COMET assay" has been developed to measure various types of lesions in chromosomal DNA, such as alkali-labile sites, DNA–DNA and DNA–protein cross-links, and single- and double-strand breaks (SSBs and DSBs) [4, 5]. The COMET assay became popular in genetic toxicology studies during recent years.

These various endpoints have been used to assess the damage (*in vitro* in cultivated cells, and *in vivo* in experimental animals and humans) induced by genotoxic agents. Test strategies are formulated by various agencies and followed by researchers worldwide. Cytogenetic methods to test possible mutagenic activities of various agents are discussed in detail below.

4.2
Cell Cycle

Cultured somatic mammalian cells *in vitro* are generally cycling in the presence of culture medium containing appropriate growth factors and antibiotics added to the culture medium. These cells pass through different phases of cell cycle, i.e. G_1 phase (the longest phase of the cell cycle with variable duration of up to about 24 h), S phase (DNA synthesis phase, duration about 7 h), G_2 phase (a short lag phase, duration about 3 h) and M phase (mitotic division, duration about 0.5–1 h). Germ cells are characterized by meiotic divisions, but these are not discussed here, though germ cells have also been analyzed for CAs.

The Role of Evidence in Risk Characterization: Making Sense of Conflicting Data.
Edited by Peter M. Wiedemann and Holger Schütz
Copyright © 2008 WILEY-VCH Verlag GmbH & Co. KGaA, Weinheim
ISBN: 978-3-527-32048-6

Human peripheral blood lymphocytes (HPLs), which are the most studied human cells in genetic toxicology, normally do not go through the cell cycle and proliferate. An exogenous mitogenic stimulant, generally phytohemagglutinin (PHA), is required to be added to the culture medium to induce cell division. In peripheral blood HPLs are in a nonproliferating phase of the cell cycle (G_0 phase). During G_0 or G_1, chromosomes in all cells contain only one chromatid with one continuous DNA molecule. Upon mitogenic stimulation HPLs leave the G_0 phase and enter into the G_1 phase, followed by the S phase during which the single DNA molecules in each chromatid are replicated to give rise to two chromatids, each containing one DNA molecule. After a short G_2 phase, the cell enters mitosis (M phase) when the two chromatids and therefore the two DNA molecules move towards opposite spindle poles of the dividing cell. Division of the cytoplasm gives rise to two daughter cells. The duration of the G_1 phase varies among different cell types while the length of S phase, which is dependent on the DNA content, is quite similar in mammalian cells. The cell cycle duration for HPLs is as follows: G_0–G_1 transition phase about 12 h after PHA stimulation; S phase starts about 24 h after PHA stimulation followed by a short G_2 phase of about 3 h. The first mitoses (M_1) occur at about 40 h and metaphases are prepared at about 48 h. The cell cycle parameters given are mean values. The duration of the cell cycle in HPLs is dependent on the type of mitogen, the culture medium and the individual blood sample [6, 7]. Normally, under *in vitro* culture conditions, there are 0–5% HPLs in their second mitotic division (M_2) after 48 h culture time, but this is not standard for HPLs of all individuals. HPLs from some individuals proliferate faster, in rare cases even more than 50% M_2 cells were found after 48 h [7]. In all CA studies conducted *in vitro*, cells in M phase are accumulated by the addition of colchicine (or a chemical derivative of it) to the culture medium for about 4 h which disrupts the spindle formation leading to an accumulation of C metaphases. Chromosomes in C metaphases are shorter than in normal metaphases and exhibit clearly visible chromatids and centromeres. C metaphases are not followed by cell division, but eventually give rise to tetraploid cells. In the following C metaphases are called metaphases (M phases).

4.3
Test Systems

4.3.1
COMET Assay to Evaluate Primary DNA Damage

As soon as the cells are exposed to a genotoxic agent, *in vivo* or *in vitro*, primary lesions such as DNA–DNA and DNA–protein cross-links, SSBs, DSBs, and alkali labile sites are induced in DNA. Cells have an inherent capacity to repair these lesions [8]. When repair mechanisms do not restore the original structure of DNA (mis-repair), CAs and in consequence MN can be formed. SCEs are the visible result of recombination repair between the DNA of sister chromatids. Several biochemical techniques are available to determine the extent of primary DNA damage in cells. However, a simple technique, the COMET assay, has been developed and increasingly used in recent years. The laboratory protocol involves embedding single cells in agarose on microscopic slides, lysis of cellular structures and unwinding of nuclear DNA. This is followed by electrophoresis during which the induced lesions may separate from the intact DNA and driven by the electric field move away from the nuclear area to make the cell appear as a "COMET"; hence the name – COMET assay. Alkaline and neutral pH conditions used to electrophorese damaged cellular DNA should preferably lead to the release of alkali-labile lesions plus SSBs and DSBs, respectively, but it seems that alkaline conditions are the most effective to detect DNA damage [4, 5]. The COMET assay does not give a quantitative measure of DNA damage, but it indicates that damage must have been induced by a genotoxic test agent. The data will be collected as the length of the COMET tail measured in microns and/or COMET moment which is the ratio derived from the amount of DNA in the COMET tail to that in the COMET head, and compared between control cells and those exposed to a genotoxic agent. Extreme caution must be used while employing the COMET assay for cultured, continuously growing cells since the results will be considerably influenced by the presence of cells in apoptosis (programmed cell death) and those in S phase, both could mimic the damaged cells [9]. HPLs in G_0 are especially suited for performing the COMET assay because they contain no proliferating cells.

4.3.2
Chromosomal Aberrations

It is generally accepted that the ultimate lesions for the formation of CAs are DSBs. However, the exact mechanism(s) by which CAs are formed from DSBs is not yet clear [10]. Is one DSBs enough to trigger the formation of a CA or are two DSBs necessary for this? What are chromosomal breakpoints in CAs telling us about the primary lesions which gave rise to the CAs in question? It has been shown repeatedly that the intra- and inter-chromosomal distributions of breakpoints are not random. Does this tell us something about the primary lesions which led to the formation of CAs? Actually not, breakpoints of CAs are selected observed residual breakpoints (SORBs) which cannot directly be correlated with the primary lesions induced by a chromosome breaking agent [10]. Most of the DSBs that are primarily induced leave no traces in the microscopic picture of chromosomes. Nonetheless, a positive correlation has been demonstrated between CAs and carcinogenesis. CAs result from DSBs and are reliable indicators of DNA damage induced by a test agent; hence, it is not unexpected that CAs are closely associated with cancer.

CAs are generally analyzed in M phase of the cell cycle. Special experimental procedures have been developed forcing the chromosomes in cells which are not in M phase of the cell cycle to undergo premature chromosome condensation (PCC). PCC allows CA analysis of chromosomes in G_1 or G_2 phase, where they are recognizable as elements with one or two chromatids, respectively. PCC in S phase (S-PCC) exhibits double fragments of chromatin with unstained regions between them. The fragments are already replicated and the unstained areas between them are sites of DNA synthesis. S-PCC is not suited for analyses of CAs. G_1 PCC can be used to analyze

CAs, but this is not done in routine work [11]. PCC in G_2 phase (G_2-PCC) is easily inducible in human lymphocytes and is used for routine CAs analysis [12].

DSBs occur spontaneously in cells quite frequently, but most of them are repaired [13, 14]. Only a few genotoxic agents are known to induce DSBs directly which when mis-repaired lead to the formation of CAs in the same stage of the cell cycle in which they were induced. In G_0/G_1 phase of the cell cycle, mis-repaired DSBs will give rise to CAs in the single chromatid which is replicated during S phase and in the ensuing first division M phase (M_1) or in G_2-PCC will affect both chromatids at the same site. Such aberrations are classified as chromosome-type CAs, in chromatid-type CAs only one chromatid of the metaphase chromosome is aberrant. In S phase of the cell cycle, mis-repaired DSBs give rise to both chromosome and chromatid-type CAs, depending on whether DSBs were induced in replicated or not yet replicated DNA. In G_2 phase, mis-repaired DSBs lead to the exclusive formation of chromatid-type CAs. Genotoxic agents which directly induce DSBs are ionizing radiation, neutrons, high-energy heavy ions, some antibiotics such as neocarcinostatin and bleomycin, and endonucleases such as restriction endonucleases. These agents are referred to as S phase independently CA inducing agents. Most DNA-damaging agents induce lesions other than DSBs and like DSBs most of these will be repaired [8, 15]. When un-repaired damage other than DSBs is present during S phase it may be transformed to DSBs at the replicating sites and these may then give rise to chromatid-type aberrations in M_1 phase of the cell cycle. When damage other than DSBs is induced in G_2 phase of the cell cycle, this does not lead to CAs in M_1. With respect to their CAs inducing activity, these types of genotoxic agents are classified as S phase dependently CA inducing agents.

The structures as well as the numbers of chromosomes are best discernable in M phase when both chromatids of the chromosomes are condensed and clearly distinguishable. It is important to accumulate as many cells in M phase as possible to examine the recommended number of cells to be scored; this is achieved by exposure to colchicine (see Section 4.2). Since the early 1960s, CAs have been analyzed in chromosomes stained with Giemsa stain. Giemsa staining results in uniformly stained chromosomes, or after special pre-treatment to banded chromosomes such as centromeric (C)-, Giemsa (G)- and reverse (R)-banding. The more recently developed method to paint specific chromosomal areas such as centromeres, single chromosomes or all chromosomes using molecular probes by fluorescent *in situ* hybridization (FISH) is highly effective and uncovered very complex CAs [16, 17]. These techniques are expensive and time consuming, and may not be suitable for large-scale investigations.

It has been shown repeatedly that specific types of CAs are associated with cancer [18]. Analyses of spontaneous frequencies of CAs in peripheral blood lymphocytes of human populations revealed that people with CA frequencies in the higher tertile of the frequency distribution of CAs in a given population have a significantly higher cancer risk than the rest of the population [19]. This and the close association between CA-inducing agents and their carcinogenic activities in experimental animals and in exposed people show the utmost importance of CAs as indicators for genotoxic as well as carcinogenic activities.

4.3.3
Micronuclei

The existence of MN has been known for many years. At least two mechanisms of action of genotoxic agents contribute to the formation of MN: (i) broken pieces of chromosomes (chromosome breaking or clastogenic action) and/or (ii) unequal segregation of whole chromosomes during cell division due to spindle disruption (aneugenic action). The basic premise was that larger MN are caused by spindle disruption, while smaller MN would consist of one or more chromosomal fragments. However, there is a significant amount of overlap between these two classes of MN with the result that it is impossible to know for certain how a single MN is formed. Recent molecular technologies have paved the way to elucidate either or both of these mechanisms (clastogenic and aneugenic action). This is to employ centromeric probes to hybridize with centromeric DNA in MN or the use of CREST antibodies to hybridize MN with centromeric proteins. This allows the distinction whether or not a MN contains a centromere, in which case it should contain a whole chromosome (aneugenic effect), and if not the MN is formed from broken piece(s) of chromosomes (clastogenic action). Nonetheless, one cannot rule out the possibility of the presence of chromosomal fragments in MN with centromeres although it can be expected that in most of the cases MN with centromeres will contain whole chromosomes. MN are usually analyzed in cells exposed to a test agent and grown in the presence of cytochalasin B (CytB) which blocks cellular, but not nuclear division. This leads to binucleate cells in which MN can be easily analyzed. In analyses with HPLs, CytB is added at 44 h after the initiation of cultures and cells are fixed at a culture time of 72 h [20].

4.3.4
Sister Chromatid Exchanges

SCEs are cytological manifestations of a recombination repair mechanism of DNA damage in S phase resulting in interchanges between the two chromatids of the same chromosomes at apparently homologous loci. Possibly this happens at the replication fork itself [21, 22]. In general, SCEs are more sensitive indicators of genotoxicity as compared with CAs and MN. The reason for this is that the spontaneous frequencies of SCEs are quite high (about 5 SCEs/cell) and less cells have to be analyzed to proof a significant elevation of this end point by a given test agent. SCE evaluation can be only carried out in chromosomes of cells that are going through their second M phase (M_2). The development of the fluorescence plus Giemsa (FPG) technique [23] has aided in the extensive use of this technique during the late 1970s. It is generally believed that increased incidence of SCEs does not indicate real adverse effects, but points to the fact that cells were exposed to a genotoxic agent and were able to repair DNA damage. SCEs can also occur in association with CAs induced in G_1 (false SCEs as contrasted to true SCEs) [22], but this is not of importance in genetic toxicology testing [24].

4.3.5
Other Assay Systems and Endpoints

Alterations in DNA sequences are evaluated in tests other than COMETs, CAs, MN and SCEs. An extensively used assay is the Ames test in which specific mutant strains of the bacterium *Salmonella typhimurium* are exposed to genotoxic agents and then analyzed for reverse mutations of the genes tested. Generally, the assay is performed in the presence of microsomal metabolizing enzymes prepared from rat liver to determine whether or not the test agent requires metabolic activation to exert its genotoxic effect. Similar tests for gene mutations are available using *Escherichia coli*, yeast and mammalian cells. Metabolizing enzymes extracted from rat liver are also used in techniques for the evaluation of CAs, MN and SCEs in cultured mammalian cells including HPL *in vitro* [25]. All types of mutations (such as gene and chromosome mutations as well as aneuploidies) can be analyzed in *Drosophila melanogaster*.

4.4
Methodological Aspects

It is extremely important to examine the first post-treatment M phase (M_1) for CA analyses. During the following cell cycles (M_2, M_3+) there is a considerable selection against aberrant cells which do not allow making reliable quantitative decisions. In addition, the aberration patterns change and it is no longer possible to decide which types of CAs have been primarily induced by the genotoxic agent in question. Dicentric chromosomes induced in G_0/G_1 by an S phase-independently CA-inducing agent will be eliminated during successive cell divisions. In contrast, reciprocal translocations will survive longer and can still be found in later divisions. Some types of chromatid exchanges induced by S-dependently CA-inducing agents will give rise to translocations and others to dicentric chromosomes [26]. Obviously this does not allow quantitative and qualitative relationships to be derived between a test agent and the CAs induced by it. Controlling the cell cycle can be done by allowing the cells to incorporate 5-bromo-2′-deoxyuridine (BrdU) instead of thymidine (T). After one cell cycle one strand of the DNA of both chromatids is substituted with the bromine (B) derivative (TB–TB). In M phase chromosomes in the second cell cycle (M_2 phase) DNA in chromatids will be TB–BB and these can be differentially stained with the FPG technique (TB chromatids are darkly stained and BB chromatids are lightly stained). In M phase chromosomes in the third and following cell cycles ($M3+$ phase), lightly stained (BB–BB) and differentially stained (TB–BB) chromosomes occur in the same metaphases. In TB–BB chromosomes, SCEs can be recognized as switches between darkly and lightly stained chromatids. This methodology can be used to recognize whether metaphases represent first (M_1), second (M_2) or third+ (M_3+) cell divisions. CAs should only be analyzed in uniformly dark stained M_1 metaphases. If it can be proven that there are not more than 5% of M_2 phase cells, CAs can be analyzed in cells not stained differentially, because 5% M_2 cells cannot very much influence the result. In differentially stained M_2 phase cells SCEs can be analyzed, but it should be mentioned that an interaction of the test agent with the incorporated BrdU and/or with BrdU in the medium may influence the result of an SCE test [22].

4.4.1
In Vitro Studies

4.4.1.1 CAs in HPLs

HPLs are well suited for cytogenetic analyses. In the peripheral blood generally more than 95% cells are in a resting phase of the cell cycle (G_0 phase). In culture HPLs can be stimulated to enter the mitotic cell cycle by adding a mitogenic compound to the culture medium, generally PHA. With respect to cell cycle control HPLs are an excellent test system and allow performing highly controllable experiments. It should be mentioned here that all cytogenetic investigations must be conducted following Good Laboratory Practice (GLP). It is imperative to include sham and/or untreated control cells as well as positive control cells that are treated with a known genotoxic agent, especially when a test compound has only slight or borderline effects on the genetic endpoint analyzed. At least three independent experiments have to be performed in the case of HPLs with the blood from the same person to be able to check the variability between experiments. Experiments in which several cultures are set up at the same time and exposed in the same way are not independent, but parallel experiments. Independent experiments allow calculating standard errors of the means which should be given in respective publications. The number of cells to be analyzed depends on the endpoint to be looked at. Variability in the CA frequencies of HPLs from different donors and interlaboratory variability in scoring CAs is well documented in the literature [27, 28]. These findings emphasize the need for a substantial amount of cells (up to 1000) to be examined for getting reliable results.

4.4.1.2 CAs in Fibroblasts

Fibroblasts of different origin are also used in cytogenetic experiments to assess genetic damage. One difficulty of using fibroblasts is to properly control the cell cycle. Primary fibroblasts which can be set up from human tissues or from tissues of animals have stable karyotypes, but after prolonged culture period the cells might acquire other characteristics such as changes in chromosome structure and numbers apart from going through natural senescence. Some cells may eventually be able to grow permanently *in vitro*. Permanently dividing cultured cell lines have unstable karyotypes, but they are quite useful because of their effective and unlimited growth characteristics. Permanently growing cells, such as Chinese hamster ovary cells, are widely used in cytogenetic testing. Cell lines from various tumours are also applied in cytogenetic experiments.

4.4.1.3 MN

The frequencies of MN in HPLs are highly variable and dependent on various factors, such as culture conditions, scoring criteria gender and age of the person from whom the analyzed HPLs were taken [29]. The usually recommended number of cells analyzed for MN is 1000–2000 binucleate cells.

4.4.1.4 SCEs
It is demonstrated in the literature that the incidence of SCEs depends on various factors, such as source of lymphocytes, concentration of BrdU, type of serum in the culture medium and gender. The usually recommended number of M_2 phase cells analyzed for SCEs is 50. The normal incidence of SCEs in lymphocytes is about 5 SCEs/cell.

4.4.1.5 Metabolic Activation
Several chemical mutagens are only genotoxic when they are metabolically activated. It is therefore important to test all suspected genotoxic agents both in the presence as well as in the absence of metabolic enzymes extracted from rat liver [25].

4.4.2
In Vivo Studies

4.4.2.1 Mammals
The cytogenetic endpoints described above, i.e. CAs, MN and SCEs, can also be studied under *in vivo* conditions. In order to do this, cells are to be taken from tissues that are mitotically active or that can be stimulated *in vitro* to undergo mitotic divisions in culture. In laboratory rodents such as rats and mice, the bone marrow is a widely used tissue for genotoxicity analyses. After exposure of animals with a test agent, cells from bone marrow are taken and prepared for analyzing CAs or MN. Previous application of BrdU allows analyzing SCEs. Apart from bone marrow, other tissues can also be used to study MN [30]. In case a test compound does not reach the target tissue, the result of an *in vivo* cytogenetic study will be negative.

4.4.2.2 Humans
HPLs form an excellent cell system to study cytogenetic effects of human exposure *in vivo*. HPLs of exposed and unexposed persons are cultured *in vitro*, and M_1 phase (for CAs), M_2 phase and binucleate cells (for SCEs and MN, respectively) are analyzed. Significantly increased CAs, MN and SCEs in cells from exposed as compared with those from unexposed subjects would indicate that the test agent being investigated is mutagenic and carcinogenic. Individuals in the control group must be as similar as possible to the individuals in the exposed group, except the exposure. The exposure condition should be well characterized in order to ensure that an elevation of the genetic endpoint tested is really an effect of an exposure in question and not of something else. The numbers of individuals in control and exposed groups as well as the numbers of cells examined in each individual must be large enough to confirm the genotoxic effect of the exposure [31]. Exposures of people to chromosome damaging effects can be acute or chronic. Acute exposure occurs in radiation accidents and leads to typical CAs such as dicentric chromosomes and translocations. The frequencies of these CA types after a given radiation dose applied *in vivo* or *in vitro*

are quite similar and therefore CA frequencies obtained *in vivo* can be used to retrospectively estimate the radiation dose ("biological dosimetry") [32]. In most of the cases, exposures in man are chronic (such as in occupational exposure situations) and here too HPLs are a good system to determine genotoxic effects. The lifespan of HPLs is quite long (half-life about 3 years) and they can accumulate damage by exposure over years. Since HPLs are in a presynthetic stage of the cell cycle (G_0 stage) exposure to agents with an S-dependently CA-inducing activity should lead to chromatid-type aberrations when cells are stimulated to undergo mitoses *in vitro*. This expectation is generally not fulfilled in that apart from chromatid-type aberrations, chromosome-type aberrations are also found to be elevated, how is this possible? HPLs exposed to an S-dependent chemical mutagen *in vitro* can be held in complete medium without PHA. Under this condition HPLs are kept alive in culture in the G_0 stage of the cell cycle, but are not stimulated to undergo mitotic divisions. Mitotic activity only occurs when PHA is applied after some time in culture (liquid holding experiments) [33]. Under such conditions, both chromosome- and chromatid-type aberrations can be identified although from the mutagen applied only chromatid-type aberrations are expected to be formed when the cells pass through the cell cycle. An explanation of this could be that under liquid holding conditions lesions induced by the mutagen may be transformed to DSBs spontaneously or by enzymatic activities, and such DSBs could eventually lead to the formation of chromosome-type aberrations by mis-repair. *In vivo* chronic exposure to mutagens in humans is a liquid holding-like situation for a long time and this may be an explanation for the occurrence of both chromosome- and chromatid-type CAs after exposure to mutagens with an S-dependent mechanism of action. After exposure to ionizing radiation analyses of HPLs reveal mainly chromosome-type aberrations independent on the type of exposure (chronic or acute). Typical aberrations after exposure to ionizing radiation *in vivo* in man are dicentric chromosomes. As discussed above, dicentric chromosomes also occur after *in vivo* exposure to S-dependently acting mutagenic agents and are therefore not typical for an exposure to ionizing radiation.

In vivo test systems guarantee that the test agent is investigated under metabolic conditions since the liver naturally transforms an inactive agent to a genotoxic agent. A typical example is alcohol. Alcoholics have significantly elevated chromosome- and chromatid-type CAs in their peripheral lymphocytes, and epidemiological evidence confirms that alcohol is carcinogenic in man. *In vitro* experiments show that acetaldehyde – the first metabolite of ethanol – but not alcohol itself induces CAs and SCEs [34].

4.5
GLP

Elevated indices of CAs, MN and SCEs are the result of damage to chromosomal DNA, and the ultimate lesions responsible for their formation are DSBs. These can be induced directly or by transformation of lesions other than DSBs during S phase of the cell cycle. In order to confirm the genotoxic effect of an unknown agent, careful consideration and attention must be paid with respect to the experimental protocols and procedures used in the laboratory.

(1) All investigations MUST be conducted "blind".

(2) All studies MUST include positive and sham/untreated/unexposed controls, i.e. cells treated with known genotoxic agent(s) and sham and/or untreated controls, respectively.

(3) The dose–effect relationship of the test agent should be investigated; therefore at least three increasing doses of the test agent must be tested. A positive correlation of the results with the dose of the test agent should be observed.

(4) Evaluation of more than one genotoxicity endpoint is recommended. The data obtained from all endpoints must point in the same direction. There should be either a significant increase in exposed cells as compared with those in control cells (positive direction) or no significant difference between exposed and control cells (negative direction). If the results from two endpoints are discordant, e.g. CAs being negative and MN being positive, further investigations must be carried out to identify whether the MN contain whole chromosomes and are therefore the result of aneugenic activities. If the treatments are carried out in a presynthetic stage of the cell cycle, chromosome-type aberrations in M_1 phase indicate that the agent in question is able to induce DSBs directly. A concurrent SCE test would be negative or at best borderline positive in such a test schedule. Chromatid-type aberrations would indicate that the test agent induces lesions other than DSBs, an SCE test would be positive in this case.

(5) In case a test agent gives negative results with all endpoints it could be that the agent is genotoxic only upon metabolic activation. This should be confirmed with the addition of metabolic enzymes extracted from rat liver to the cell cultures.

(6) Apart from the considerable variability of cytogenetic tests it is absolutely necessary to perform at least three independent experiments. This is extremely important and mandatory with agents exhibiting borderline genotoxic effects.

(7) The COMET assay indicates DNA damage. However, the test by itself is not enough to draw the conclusion that the test agent is genotoxic; it has to be combined with other tests.

(8) The SCE assay indicates DNA damage. Similar to the COMET assay, it does not allow drawing conclusions whether a test agent is genotoxic and, hence, carcinogenic. The substitution of chromosomal DNA with BrdU and/or BrdU in the medium may influence the outcome of the test.

(9) *In vivo* tests with directly DSB-inducing agents such as X-rays should give similar results as *in vitro* tests performed with the same doses. This must not

be the case with chemical agents which induce various other types of DNA damage and which may only be active after metabolic transformation.

(10) If several tissues/organs are examined *in vivo* for genotoxic effects, the test agent might not have reached the target tissue. This must be kept in mind when the data are negative.

(11) A test agent turning out to be negative *in vitro* could be positive *in vivo* when the test agent is genotoxic only after metabolic activation.

(12) *In vivo* studies with HPLs of persons exposed to a suspected or known mutagenic agent should give positive results if the agent or its metabolized compound exerts genotoxic effects.

(13) It is useful to include both *in vitro* and *in vivo* investigations using multiple endpoints to confirm whether or not a test agent is genotoxic.

4.6 Evidence Characterization and Interpretation of Genetox Results

4.6.1 Interpretation of Data from One Endpoint

(1) *COMET positive*. If the test is performed immediately after treatment of the cells with a test agent, a positive result would indicate that the test compound is capable of inducing primary DNA damage. However, the data do not provide information whether such damage could lead to SCEs by repair or to CAs by mis-repair and in consequence to MN. Positive data obtained from the COMET assay alone do not allow drawing conclusions, but they indicate that the test agent is able to induce lesions in DNA.

(2) *CA positive*. The test compound is capable of inducing primary DNA damage which by mis-repair leads to the formation of CAs. CAs are the most reliable endpoint to predict the mutagenic and carcinogenic potential of a test agent.

(3) *MN positive*. The mechanism by which the MN are formed must be investigated using centromere staining. This is to differentiate between the clastogenic and aneugenic effect of the test agent. The data must still be interpreted with caution since broken chromosomes containing centromeres when occurring in MN would also appear like whole chromosomes with positive centromere staining. Nevertheless, in cases where the majority of MN contains centromeres an aneugenic action (spindle disruption) by the test agent is highly probable.

(4) *SCE positive*. Investigators must keep in mind that the incidence of SCEs is positively correlated with the concentration of BrdU in the culture medium and in consequence in chromosomal DNA. The SCE test alone indicates that the test

compound is capable of inducing lesions in the DNA or that there is an interaction of the test compound with the BrdU in chromosomal DNA and/or in the medium. Nonetheless, no adverse effect on human health is reported in the literature as a consequence of increased SCEs. The SCE test is an indicator test showing that the test agent is able to damage DNA.

Except for positive CAs, all other single endpoint studies require additional investigations to draw meaningful conclusions; hence, it became customary and mandatory to use a battery of tests in genetic toxicology investigations since no single test or endpoint provides confirmatory evidence on the genotoxic potential of a test agent. It is generally believed that the data from multiple endpoints follow in the same direction, i.e. either positive (significantly increased damage) or negative (absence of such increase in damage) in exposed as compared to unexposed cells or individuals. Nevertheless, different combinations of positive and negative results can be obtained when multiple endpoints are investigated. Below are only some such examples where the data need to be interpreted with caution.

4.6.2
Interpretation of Data from Four Endpoints

(1) *COMET, CAs and MN negative, SCEs positive*. Positive results from the SCE test alone could indicate that the test compound is capable of interacting with the DNA or may interact with BrdU in chromosomal DNA and/or in the culture medium. Also, since the other endpoints are negative, the dose(s) of the test compound used to treat the cells could have been too low to induce increased frequencies of CAs and MN. Nonetheless, the COMET assay should have been positive even with a low dose of test agent. Additional investigations are required to draw firm conclusions on the genotoxic potential of the test agent.

4.6.3
Interpretation of Data from Three Endpoints

(1) *COMET positive, CA and MN negative*. If the COMET assay is performed immediately after treatment of the cells with a test agent, the COMET data could be positive while the primary damage might have been efficiently repaired, leading to negative effects with respect to CAs and MN.

(2) *COMET negative, CA and MN positive*. The COMET assay measures primary DNA damage which is often repaired completely within a few hours by the repair mechanisms of the cell. If the COMET assay is performed several hours after the induction of DNA damage, the data from the COMET assay could be negative due to the efficient repair of induced damage. Mis-repaired damage could lead to increased incidence of CAs and MN in the ensuing first and second cell cycle (M_1 and M_2 cells), and thus yield positive results for CAs and MN. With respect to MN, a potential aneugenic action (disruption of the spindle) of the test agent needs to be considered.

4.6.4
Interpretation of Data from Two Endpoints

(1) *COMET negative and CA positive*. The COMET assay measures primary DNA damage, which is often repaired completely within a few hours by the inherent repair mechanisms of the cell. If the COMET assay is performed several hours after the initiation of DNA damage, the data from the COMET assay could be negative due to the efficient repair of induced damage. Mis-repaired damage could lead to the formation of CAs in the ensuing M_1 cells and this would explain the positive data of the CAs test.

(2) *COMET negative and MN positive*. The COMET assay measures primary DNA damage that is often repaired completely within a few hours by the inherent repair mechanisms of the cell. If the COMET assay is performed several hours after the initiation of DNA damage, the data from the COMET assay could be negative due to the efficient repair of induced damage. Mis-repaired damage could lead to the formation of MN in M_2 cells. The data could also be interpreted as an absence of induction of primary DNA damage resulting in a negative COMET assay. The MN could be an aneugenic (spindle disruption) effect of the test agent. Additional studies must be performed to determine the presence of centromeres in the MN. However, this interpretation could still be a problem since the presence of broken chromosomes with centromeres in MN could be interpreted as aneugenic effect of the test agent.

(3) *COMET negative and SCE positive*. The data could indicate that the primary DNA damage that is induced is efficiently repaired and the positive SCE data may or may not indicate a real genotoxic effect of the test agent. SCEs could be the result of an interaction of the test agent with BrdU in DNA and/or in the medium. Additional investigations are required.

(4) *CA positive and MN negative*. Additional investigations are required since the data are inconclusive.

(5) *CA negative and MN positive*. The test agent might have an aneugenic effect (spindle disruption) that has to be checked by analyzing MN for the presence of centromeres.

(6) *CA negative and SCE positive*. The SCE test is more sensitive than the CA test, because the spontaneous frequencies of SCEs are much higher than those of CAs. The positive SCE test indicates damage in DNA. The concentration of the test agent used to treat the cells might have been so low that only elevations of SCEs but not of CAs have been proven. The negative CA test could result from too few mitoses being analyzed. In addition, an interaction of the test compound with BrdU in DNA and/or the medium has to be taken into consideration.

(7) *MN negative and SCE positive*. The SCE test is more sensitive than the MN test. The positive SCE test indicates damage in DNA. The concentration of the test agent used to treat the cells might have been so low that SCEs but not MN have

been proven. An interaction of the test agent with BrdU in DNA and/or the medium could be responsible for the positive SCEs test.

4.7
Genetox Studies with Electromagnetic Fields

During recent decades, a large number of investigations have been conducted using experimental animals, cultured rodent and human cells as well as freshly collected HPLs to determine the genotoxic potential of *in vivo* and/or *in vitro* exposure to extremely low-frequency (ELF) and radiofrequency (RF) electromagnetic fields (EMFs) (for reviews, see Refs [35–39]). Vijayalaxmi and Obe [37, 38] reviewed the data from 53 and 63 publications evaluating the cytogenetic effects of RF and ELF exposures, respectively. The genetic endpoints examined were COMET assay, CAs, MN and SCEs. In both data sets, about 50% of the papers reported negative (ELF = 46%; RF = 58%) and 20% of the papers reported positive results (ELF = 22%; RF = 23%). In 32% of ELF and in 19% of RF studies the data were inconclusive. Hence, from the overall data, it is not possible to categorize ELF and RF radiation as "potential" genotoxic/mutagenic agents. If ELF and RF radiation would be mutagenic the data should provide unequivocal positive results as is the case of ionizing radiation and other mutagens. Nevertheless, the positive cytogenetic data that were reported with ELF and RF radiation have created controversy in the scientific community and in the public mind. The potential reasons for the positive findings are several-fold and are described in detail [37, 38]. They include: (i) experimental and/or analytical flaws, (ii) variability of the data have not been properly taken into account, (iii) only one genetox endpoint was investigated which did not indicate the possible spectrum of cytogenetic effects, and (iv) when multiple endpoints were investigated, the observations did not correlate (point in the same direction as positive or negative).

These controversial observations reported in nonionizing radiation literature can only be resolved in carefully planned experimental studies that take into account the variability range to be expected in cytogenetic tests. One important source of variability is scoring for cytogenetic effects. In order to come to a decisive conclusion of whether ELF and RF radiation have a genotoxic potential it is important to conduct international collaborative investigations involving scientists with expertise in cytogenetics. Exposures and slide preparation should be done in one laboratory under highly controlled conditions. Analyses are performed in the different participating laboratories. An example of such an extensive study is that of Lloyd *et al.* [27] in which CAs induced by very low doses of ionizing radiation in human lymphocytes *in vitro* were analyzed by experienced cytogeneticists in independent laboratories in several countries across the world. Lymphocytes from 24 donors were used and about 300 000 metaphases were analyzed. This analysis uncovered various sources of variability such as differences in scoring results between participating laboratories irrespective of the fact that only experienced cytogeneticists participated in the study. A reason for the interlaboratory variability of scoring results could not be found. The extensive investigation of Lloyd *et al.* [27] highlighted the importance of variability, especially when a borderline increase in genotoxicity was observed in lymphocytes exposed to very low doses of ionizing radiation – this could very well be the case with exposures to nonionizing radiation, and ELF and RF radiation.

References

1 Evans, H.J. (1976) Cytological methods for detecting chemical mutagens. In *Chemical Mutagens Principles and Methods for Their Detection*, (ed. A. Hollaender), pp. 1–29, Plenum Press, New York.

2 Müller, W.-U. and Streffer, C. (1994) Micronucleus assays. In *Advances in Mutagenesis Research*, vol. 5 (ed. G. Obe), pp. 4–134, Springer-Verlag, Heidelberg.

3 Obe, G. and Natarajan, A.T. (1993) Mutagenicity tests with cultured mammalian cells: cytogenetic assays. In *Handbook of Hazardous Materials*, (ed. M. Corn), pp. 453–461 Academic Press, New York.

4 Müller, W.-U., Ciborovius, J., Bauch, T., Johannes, C., Schunck, C., Mallek, U., Böcker, W., Obe, G. and Streffer, C. (2004) Analysis of the action of the restriction endonuclease *Alu*I using three different comet assay protocols, *Strahlentherapie und Onkologie*, **180**, 655–664.

5 Speit, G. and Hartmann, A. (2006) The comet assay: a sensitive genotoxicity test for the detection of DNA damage and repair, *Methods in Molecular Biology*, **314**, 275–286.

6 Natarajan, A.T. and Obe, G. (1982) Mutagenicity testing with cultured mammalian cells: cytogenetic assays. In *Mutagenicity New Horizons in Genetic Toxicology*, (ed. J.A. Heddle), pp. 171–213, Academic Press, New York.

7 Obe, G. and Beek, B. (1982) The human leukocyte test system. In *Chemical Mutagens – Principles and Methods for Their Detection*, vol. 7 (eds F.J. de Serres and A. Hollaender), pp. 337–400, Plenum Press, New York.

8 Friedberg, E.C., Walker, G.C. and Siede, W., (1995) *DNA Repair and Mutagenesis*, ASM Press, Washington, DC.

9 Vijayalaxmi, McNamee, J.P. and Scarfi, M.R. (2006) Letter to the Editor: comments on: "DNA strand breaks" by Diem *et al.* [Mutat. Res. 583 (2005) 178–183] and Ivancsits *et al.* [Mutat. Res. 583 (2005) 184–188], *Mutation Research*, **603**, 104–106.

10 Obe, G., Pfeiffer, P., Savage, J.R.K., Johannes, C., Goedecke, W., Jeppesen, P., Natarajan, A.T., Martinez-Lopez, W., Folle, G.A. and Drets, M.E. (2002) Chromosomal aberrations: formation, identification and distribution, *Mutation Research*, **504**, 17–36.

11 Rao, P.N., Johnson, R.T. and Sperling, K. (Eds.) (1982) *Premature Chromosome Condensation Application in Basic, Clinical, and Mutation Research*, Academic Press, New York.

12 Gotoh, E., Asakawa, Y. and Kosaka, H. (1995) Inhibition of protein serine/threonine phosphatases directly induces premature chromosome condensation in mammalian somatic cells, *Biomedical Research*, **16**, 63–68.

13 Pfeiffer, P., Goedecke, W. and Obe, G. (2000) Mechanisms of DNA double-strand break repair and their potential to induce chromosomal aberrations, *Mutagenesis*, **15**, 289–302.

14 Vilenchik, M.M. and Knudson, A.G. (2003) Endogenous DNA double-strand breaks: Production, fidelity of repair, and induction of cancer, *Proceedings of the National Academy of Sciences of the United States of America*, **100**, 12871–12876.

15 Singer, B. and Grunberger, D. (1983) *Molecular Biology of Mutagens and Carcinogens*, Plenum Press, New York.

16 Johannes, C., Horstmann, M., Durante, M., Chudoba, I. and Obe, G. (2004) Chromosome intrachanges and interchanges detected by multicolour banding in lymphocytes: searching for clastogen signatures in the human genome, *Radiation Research*, **161**, 540–548.

17 Rautenstrauß, B.W. and Lier, T. (eds.) (2002) *FISH Technology*, Springer, Berlin.

18 Mitelman, F., Mertens, F. and Johansson, B. (1997) A breakpoint map of recurrent chromosomal rearrangements in human neoplasia, *Nature Genetics*, **15**, 417–474.

19 Bonassi, S., Znaor, A., Norppa, H. and Hagmar, L. (2004) Chromosomal aberrations and risk of cancer in humans: an epidemiologic perspective, *Cytogenetic and Genome Research*, **104**, 376–382.

20 Fenech, M. and Crott, J.W. (2002) Micronuclei, nucleoplasmic bridges and nuclear buds induced in folic acid deficient human lymphocytes – evidence for breakage–fusion–bridge cycles in the cytokinesis-block micronucleus assay, *Mutation Research*, **504**, 131–136.

21 Painter, R.B. (1980) A replication model for sister-chromatid exchange, *Mutation Research*, **70**, 337–341.

22 Wojcik, A., Bruckmann, E. and Obe, G. (2004) Insights into the mechanisms of sister chromatid exchange formation, *Cytogenetic and Genome Research*, **104**, 304–309.

23 Perry, P. and Wolff, S. (1974) New giemsa method for differential staining of sister chromatids, *Nature*, **251**, 156–158.

24 Tice, R.R. and Hollaender, A. (eds) (1984) *Sister Chromatid Exchanges, Part A, The Nature of SCEs, Part B, Genetic Toxicology and Human Studies*, Plenum Press, New York.

25 Madle, S. and Obe, G. (1980) Methods to analyze the mutagenicity of indirect mutagens/carcinogens in eukaryotic cells, *Human Genetics*, **56**, 7–20.

26 Marshall, R. and Obe, G., (1998) Application of chromosome painting to clastogenicity testing *in vitro*, *Environmental and Molecular Mutagenesis*, **32**, 212–222.

27 Lloyd, D.C., Edwards, A.A., Leonard, A., Deknudt, G.L., Verschaeve, L., Natarajan, A.T., Darroudi, F., Obe, G., Palitti, F., Tanzarella, C. and Tawn, E.J., (1992) Chromosomal aberrations in human lymphocytes induced *in vitro* by very low doses of X-rays, *International Journal of Radiation Biology*, **61**, 335–343.

28 Obe, G., Fender, H. and Wolf, G. (2001) Biological Monitoring mit zytogenetischen Methoden. In *Biological Monitoring*, (ed. J. Angerer), pp. 113–123, Wiley-VCH Weinheim.

29 Bonassi, S., Fenech, M., Lando, C., Lin, Y., Ceppi, M., Chang, W.P., Holland, N., Kirsch-Volders, M., Zeiger, E., Ban, S., Barale, R., Bigatti, M.P., Bolognesi, C., Jia, C., Di Giorgio, M., Ferguson, L.R., Fucic, A., Lima, O.G., Hrelia, P., Krishnaja, A.P., Lee, T.-K., Migliore, L., Mikhalevich, L., Mirkova, E., Mosesso, P., Müller, W.-U., Odagiri, Y., Scarfi, M.R., Szabova, L., Vorobtsova, I., Vral, A. and Zijno, A. (2001) Human MicroNucleus Project: international database comparison for results with the cytokinesis-block micronucleus assay in human lymphocytes: I. Effect of laboratory protocol, scoring criteria, and host factors on the frequency of micronuclei, *Environmental and Molecular Mutagenesis*, **37**, 31–45.

30 Suzuki, H., Shirotori, T. and Hayashi, M. (2004) A liver micronucleus assay using young rats exposed to diethylnitrosamine: methodological establishment and evaluation, *Cytogenetic and Genome Research*, **104**, 299–303.

31 Albertini, R.J., Anderson, D., Douglas, G.R., Hagmar, L., Hemminki, K., Merlo, F., Natarajan, A.T., Norppa, H., Shuker, D.E.G., Tice, R., Waters, M.D. and Aitio, A. (2000) IPCS guidelines for the monitoring of genotoxic effects of carcinogens in humans, *Mutation Research*, **463**, 111–172.

32 Wojcik, A., Gregoire, E., Hayata, I., Roy, L., Sommer, S., Stephan, G. and Voisin, P. (2004) Cytogenetic damage in lymphocytes for the purpose of dose reconstruction: a review of three recent radiation accidents, *Cytogenetic and Genome Research*, **104**, 200–205.

33 Kalweit, S. and Obe, G. (1984) Liquid-holding experiments with human peripheral lymphocytes. II. Experiments with trenimon and 1-beta-D-arabinosylcytosine, *Mutation Research*, **128**, 59–64.

34 IARC (1988) *IARC Monographs on the Evaluation of Carcinogenic Risks to Humans, Alcohol Drinking, Volume* **44**, International Agency for Research on Cancer Lyon

35 Meltz, M.L. (2003) Radiofrequency exposure and mammalian cell toxicity, genotoxicity, and transformation, *Bioelectromagnetics*, **24** (Supplement 6), S196–S213.

36 Moulder, J.E., Foster, K.R., Erdreich, L.S. and McNamee, J.P. (2005) Mobile phones, mobile phone base stations and cancer: a review, *International Journal of Radiation Biology*, **81**, 189–203.

37 Vijayalaxmi and Obe, G. (2004) Controversial cytogenetic observations in mammalian somatic cells exposed to radiofrequency radiation, *Radiation Research*, **162**, 481–496.

38 Vijayalaxmi and Obe, G. (2005) Controversial cytogenetic observations in mammalian somatic cells exposed to extremely low frequency electromagnetic radiation: a review and future research recommendations, *Bioelectromagnetics*, **26**, 412–430.

39 Verschaeve, L. (2005) Genetic effects of radiofrequency radiation (RFR), *Toxicology and Applied Pharmacology*, **207**, 336–341.

5
Animal Studies
Alexander Lerchl

5.1
Introduction

Procedures for testing chemical substances in animals are standard in industrial settings, such as pharmaceutical or chemical companies. The aim is usually to estimate doses at which these substances affect the animals' physiological status, their development, cognitive behavior or fertility, respectively. Depending on the results, doses are defined that, together with safety factors, define an acceptable dose limit for exposure of humans. The substances range from known toxins to newly developed substances of unknown toxicity and medical drugs which have to be thoroughly tested before they can be used for treating diseases in humans or animals. For high-frequency nonionizing electromagnetic radiation at frequencies in the mobile phone range and at levels below thermal effects this situation is quite different for the following reasons: (i) there is so far no accepted biophysical mechanism for nonthermal effects of these fields, and (ii) and in combination with the first reason, there is no possibility to induce any "toxic" effect without thermal effects. Nevertheless, studies investigating acute or long-term effects of electromagnetic field (EMF) exposure are needed to satisfy the safety demands of the general public.

Artificial EMFs are not new. More than 100 years ago, the first AM transmitters in the kilohertz range were installed. Ever since, the transmitters have used higher and higher frequencies, and analog TV transmitters are operating in the frequency range of approximately 50 to almost 900 MHz. Despite exceptions, there was and still is no general fear concerning possible adverse health effects caused by such transmitters although their output power is often in the megawatt range. This lack of public concern can be attributed to the fact that such transmitters are often located at remote places, i.e. far away from residential areas. However, even if these transmitters are located in the middle of a town, they rarely cause protests since (i) they transmit something "good" (e.g. radio and TV), and (ii) they often were built decades ago and thus are "part of the picture". The same is true for radar stations (e.g. at airports) which operate in the gigahertz range with exceptionally high peak power. In other words, such transmitters are usually regarded as harmless.

This situation has drastically changed with the introduction of mobile communication devices. In 1992, the first Groupe Spécial Mobile (GSM) mobile phone was introduced and today the vast majority of people in industrialized nations own at least one mobile phone. Due to the fact that the output power of mobile phones is limited to up to 0.25 W on average, which is, in turn, also a consequence of the portability of such devices and their small batteries, the base stations have to be installed in close proximity to the mobile phones. These base stations, being transmitters and receivers, are therefore distributed all over the places and often close to, or in the middle of, residential areas. As the antennas of base stations have to be installed well above ground level, these stations are either mounted on top of multilevel buildings or as single, tall towers. The ubiquitous presence of antennas or towers and their unpleasant appearance are visually/esthetically annoying, to say the least.

These facts together with the first allegations about the special risk of EMFs from mobile phones and base stations caused a huge public debate about safety issues that is still ongoing. Some single studies showed that EMFs may have harmful effects on cells, organisms and even humans. Laymen are unable to discriminate between good and bad studies, moreover EMFs are confused with magnetic fields (also in the scientific literature), and thus the whole picture is very disturbing for people living in close proximity to base stations. Interestingly, several polls have shown that base stations are of much more concern than mobile phones despite the fact that exposure of mobile phone users to EMF is attributed much more to mobile phones and not to base stations.

Politicians and local authorities are in an uncomfortable situation when it comes to installations of new base stations. Depending on the legal situation in the respective country, networking companies can freely choose the locations for these base stations, provided that they find a building or a place to install the antennas or the towers. Whenever this becomes known to the neighbors, immediate opposition is inevitable. Local authorities rarely have the right to forbid such installations although they might want to protect their citizens (and voters). On the other hand, mobile phone communication is an expanding market and a source for tax money. A classical "Catch 22".

The only way out is (scientific) proof that these EMFs, especially those from base stations, are safe. In other words, evidence is needed that they do no harm. However, it is impossible to prove the nonexistence of an effect. The best one can expect therefore is a very low probability that EMFs are potentially dangerous. Since most people are afraid of cancer being caused by EMFs, many research projects on possible health effects EMF are focusing on this area. To find answers, several approaches are followed. The most fundamental one is to look for possible biophysical effects of EMFs, e.g. on the level of enzyme reactions or at biological membranes. The next level of complexity is investigations on living cells. Here, researchers can investigate the effects of EMF exposure on gene expression, DNA strand breaks and other precancerous events. The last approach not involving humans is followed by performing animal studies. All these different approaches have advantages and disadvantages, and the present chapter tries to summarize what is known about the effects of EMFs on malignancies in rodents. In particular, methodological requirements and possible pitfalls are emphasized.

5.2
Exposure Systems

In the frequency range of about 300 MHz to several gigahertz, thus including frequencies used for mobile communication (approximately 900 MHz to 3 GHz), the interesting biological effect is energy absorption of biological tissue [1]. The basic limits for whole-body exposure, expressed as specific absorption rates (SAR), are 0.08 W/kg for the general population and 0.4 W/kg for occupational exposure. For localized exposure of the general population, these limits are 2 W/kg for head and trunk, and 4 W/kg for limbs. For occupational exposure, these levels are again 5-fold higher (10 and 20 W/kg, respectively). The rational for these limits ("basic restrictions") is based solely on thermal effects, i.e. nonthermal effects played no role whatsoever when defining these limits which were adopted by many countries.

Absorption of EMFs in this frequency range by biological tissues depends on their specific absorption characteristics, which are, in turn, a consequence of their contents of EMF-absorbing molecules and structure. Thus different types of tissue (e.g. bone, muscle, fat, brain, inner organs, skin, etc.) absorb EMF energy very differently. To ensure that the SAR limits are not exceeded and due to the fact that SAR levels are complicated to estimate (see below), so-called "reference levels" for exposure were defined. These are, as electrical and magnetic field strengths, much easier to measure. Since energy absorption depends on the EMF frequency, the reference levels are frequency dependent, too. For example, the reference levels of the electrical field strength for the general population is defined as $\sqrt{(1.375 f)}$, where f is the frequency (MHz). Thus, for 2000 MHz the reference level is 52 V/m, whereas it is 35 V/m for 900 MHz.

Exposing biological specimens and humans to high-frequency EMFs is not at all trivial, mainly for two reasons. First, the system must ensure exposure levels as uniform as possible in order to prevent large variations in field strengths and thus SAR levels (hot spots, cold spots). In this respect it is important to be aware of the fact that SAR levels are proportional to the square of the electric field strength, thus variations of electrical field strength of a factor 2 cause SAR variations of a factor 4. The second reason is that the estimation of SAR values requires detailed knowledge of the three-dimensional structure of the specimen and its surroundings, if relevant (e.g. buffer solutions). For animals, serial sections are needed which provide information about the location of the different tissues. In an experiment involving Djungarian hamsters (*Phodopus sungorus*), for example, 22 different types of tissue were identified [2]. To estimate the actual SAR values and their distribution, extensive numerical modeling is required.

The situation is further complicated by the fact that animals change their location inside a cage, thus causing different EMF absorptions. These variations must be taken into account and, furthermore, they emphasize the need for uniform exposure conditions. Alternatively, animals can be restrained during exposure, e.g. by means of tubes in which the animals are transferred. This method has two major disadvantages: (i) restraining animals is a stress factor which *per se* might have an effect interfering with the possible effects of EMF exposure, and (ii) exposure is limited to a few hours per day and usually also to 5 days per week. The advantage of restraining is the low variability of SAR values which is of particular importance when certain regions of the animals (e.g. the brain) are to be exposed.

Various technical solutions for exposing animals to uniform EMFs have been developed, e.g. radial waveguides for freely moving animals [2] and Ferris wheels for restrained animals [3]. Other, less well controlled exposure systems should no longer be used. For example, using "modified" microwave ovens [4] for exposure of animals which are placed at some distance from them should no longer be accepted by scientific journals and reviewers of such papers.

Apart from the technical development of good exposure systems it must be ensured that the specimens are actually exposed correctly. This can be achieved by monitoring the strength of the electrical field inside the exposure system and storing the data on a computer for later evaluation. Of course (see below) the persons involved in the respective experiment must not know which specimens are exposed.

5.3
Sham Exposure and Cage Controls

An essential component of properly performed experiments is a sham-exposed group. This is seen today as standard, but nevertheless not as easy to implement as it may sound. "Sham" should refer to the presence of EMFs only. All other parameters should be identical. This means that at least two identical exposure systems must be installed, ideally in one room, if no space limitations exist, to ensure identical environmental parameters (temperature, humidity). Light intensities and sound levels (including ultrasound) should also be as similar as possible. In our laboratory, several experiments with freely moving mice were performed by using radial waveguides. The exposure systems (up to eight) were placed in one large room, and the amplifiers and generators were placed in such a way that the sound produced by them (although barely hearable) was almost identical in all cages.

Exposed and sham-exposed groups must not be identifiable by the persons involved in the experiments (blinded design). All cables (e.g. for EMF power), probes for temperature, etc., must be identical. The EMF output of the amplifier is to be connected by means of a "black box" with one input (from the amplifier) and at least two outputs [one or more for the exposed group(s), one for the sham-exposed group]. The code [i.e. which was the exposed group(s), and which the sham-exposed group] must be broken only after all analyses are finished, including descriptive statistical analysis.

Exposure systems are usually quite different from normal housing conditions of laboratory animals. Not only are the cages different in size, but also the surroundings, access to food and water, etc., are unusual. Therefore, cage controls are necessary which allow checking for effects of keeping animals in exposure/sham exposure units. These effects are not necessarily negative (i.e. caused by stress). In a recent investigation in mice we were able to show that cage controls lost weight during the course of experiment, whereas exposed and sham exposed animals did not [5]. This was likely a consequence of the easier accessibility of food in the exposure/sham exposure systems. In contrast, Heikkinen *et al.* reported higher weight gains in cage control mice [6]. At any rate, these studies exemplify the need for cage controls.

5.4 Replication Studies

Results of biological studies, including those investigating the possible health hazards of nonthermal EMFs, inherently show variability of the parameters investigated. The statistical analysis, therefore, can never reach a logically derived "no" when it comes to the question of whether EMFs are dangerous. The maximum to be expected is a value for the probability that EMFs are potentially dangerous. By pure chance, some experiments give "positive" results, which are an indication that EMFs have biological effects. This is especially true if many parameters are investigated in one experiment. Often, and unfortunately, this pitfall is ignored when an experiment is planned. Likewise, single positive results in a set of multiple comparisons are often overestimated, especially by nonspecialists. Nevertheless, when an experiment shows a statistically significant effect, it must be confirmed by at least one other group of investigators before it can be regarded as a firm effect.

Presumably, the study of Repacholi *et al.* [7] is well known to many scientists working in the field of nonionizing EMFs. Transgenic (Eµ-pim1) mice with a high incidence of developing lymphomas were exposed to pulsed 900 MHz EMFs for 30 min/day for up to 18 months. The survival curves and the lymphoma incidence of the exposed mice were significantly different from the control (sham-exposed) group. This result, if confirmed, would have been strong evidence for an adverse health risk of EMFs. One critical aspect of this initial study was the exposure system which caused the SAR values to vary by a factor of more than 10 (on average 0.13–1.4 W/kg) with individual values varying even to a much greater extent (0.008–4.2 W/kg) [7]. In other words, the experimental setup caused such high variations in the SAR values that the interpretation turned out to be difficult.

Some years later, an Australian group repeated the experiment with the same transgenic mice; however, this time with much better controlled exposure conditions [8]. This goal was achieved by putting animals in Perspex tubes inside the systems (Ferris wheels) during exposure. Moreover, different SAR values were tested (0.25, 1, 2 and 4 W/kg). The initial results of the Repacholi *et al.* paper were not confirmed. Also other studies showing promotion effects of EMFs on induced spontaneous or induced skin cancers [9–11] were not confirmed [6, 12–14].

Apart from studies showing an effect that needs to be re-investigated by a replication study, others are inconclusive due to low numbers of animals in the groups. They should not have been performed in the first place. It is of no use, for example, to use 16 animals per group when survival time is one of the main endpoints [15].

5.5
Interpretation of Results

Some of the early studies which identified adverse effects of EMFs violated one or more of the above-mentioned standards [9–11]. During recent years, however, the situation has considerably improved. Thus, the negative results of more recent studies altogether indicate no adverse health effects [5, 6, 8, 12–32]. The results and more details of the studies performed from 2000 to 2007 are summarized in Table 5.1.

Table 5.1 also lists sponsors that have financially supported the respective studies. It has to be mentioned that results of studies showing no effects are often viewed with skepticism by the general public when the sponsors have a supposed interest in negative findings. For the acceptance of those studies it is therefore crucial to perform them blinded until the results have been obtained, including statistical analyses when possible. When dose–response relations are of interest, however, the code must be broken earlier. Another precondition for public acceptance of negative findings is the absence of any influence of the sponsor on the interpretation and dissemination of the results. Ideally, a statement should be included in the Acknowledgements of a publication clearly showing that no such influence exists or existed at the time when the agreement between the sponsor and the respective scientific organization was reached.

5.6
Conclusions

Based on these results, one can conclude that exposure to high-frequency EMFs has no tumor-inducing or tumor-promoting effects on rats and mice, even after prolonged exposure at high exposure levels. This interpretation, however, has some limitations. (i) "Long-term" and "life-long" exposure in short-lived rodents is not directly comparable to the situation in long-lived humans. However, this argument is valid for all comparable investigations that deal with long-term exposures of short-lived animals to toxins. (ii) Exposure to "pure" EMFs, i.e. fields with fixed and well defined frequencies and modulations, is not representative for the situation when humans are exposed to a mixture of electromagnetic, and static and extremely-low frequency magnetic and electric fields with an extremely wide range of frequencies, modulations, and intensities. It is, theoretically, possible that only a mixture of these various fields may cause harmful effects. To test this possibility, however, is clearly beyond the possibilities of scientists.

Table 5.1 Parameters and results of studies from 2000 to 2007 investigating the effects of high-frequency EMFs on spontaneous or induced tumors in rats and mice.

Reference	Year	Species	Frequency	Modulation/pulse	SAR (W/kg)	Exposure per day	Exposure per week	Duration	Restrained	Blinded analysis	Power sufficient	Number	End points[a]	Result	Sponsor
24	2000	rats	836.55 MHz ± 12.5 kHz	FM	1	3 × 2 h	7 days	734 days	+	+	+	+	s, bt, cns	–	Motorola
12	2001	mice	902 MHz	GSM/Nordic Mobile Telephone	0.35/1.5	1.5 h	5 days	78 weeks	+	–	–	+	s, h, vt	–	Nokia
13	2001	mice	1.5 GHz	Time Division Multiple Access	2/0.084	90 min	5 days	19 weeks	+	–	–	+	st, ho	–	ARIB
25	2001	mice	ultra-wideband		0.01	2 min	1 days	12 weeks	–	–	–	+	s, mt, vt	–	Governmental agencies
26	2001	rats	860 MHz	CW/Pulse	1	6 h	5 days	22 months	+	–	–	+	s, bt, cns, vt	–	Motorola
27	2002	rats	900 MHz	GSM	0.017–0.07	24 h	7 days	300+ days	–	–	–	(+)[b]	s, mt	–	Telekom
8	2002	mice	898.4 MHz	GSM	0.25–4	60 min	5 days	104 weeks	+	+	–	+	s, l	–	Motorola
15	2003	rats	900 MHz	GSM	0.1–3.5	2 h	5 days	9 weeks	+	–	–	–	s, mt	inconclusive	France Telecom

(*Continued*)

Table 5.1 (*Continued*)

Reference	Year	Species	Frequency	Modulation/pulse	SAR (W/kg)	Exposure per day	Exposure per week	Duration	Restrained	Blinded analysis	Power sufficient	Number	End points[a]	Result	Sponsor
6	2003	mice	849 MHz/902.4 MHz	Digital-Advanced Mobile Phone Service/GSM	0.5	1.5 h	5 days	52 weeks	+	−	−	(+)[b]	s, c, st, m	−	Nokia
28	2003	rats	835 MHz/847 MHz	Frequency Division Multiple Access/Code Division Multiple Access	1.3	4 h	5 days	730 days	+	−	−	+	s, vt	−	Motorola
29	2003	rats	1.6 GHz	iridium	0.16 (fetus), etc	2 h	7 days	2 years	+	+	+	+	s, bt, l, mt	−	Motorola
30	2004	mice	900 MHz	GSM	0.40	24 h	7 days	10 months	−	+	+	+	s, h, l	−	Governmental agencies
14	2005	mice	849 MHz/1763 MHz	Code Division Multiple Access	0.4	2×45 min	5 days	19 weeks	−	+	−	+	st, c	−	Governmental agencies

#	Year	Animal	Frequency	System	SAR	Exposure/day	Days/week	Duration					Endpoints[a]	[b]	Funding
31	2006	rats	900 MHz	GSM	0.44, 1.33, 4.0	4 h	5 days	26 weeks	+	+	−	+	s, mt	−	Mobile Manufacturers Form, Global System for Mobile Communications Association
5	2007	mice	1960 MHz	Universal Mobile Telecommunications System	0.4	24 h	7 days	10 months	−	+	+	+	s, h, l	−	Governmental agencies
32	2007	mice	900 MHz/ 1747 MHz	GSM/ Digital Cellular System	0.4, 1.33, 4.0	2 h	5 days	2 years	+	+	−	+	s, vt, l	−	Governmental agencies, Mobile Manufacturers Form, Global System for Mobile Communications Association

[a] Endpoints (excluding weights of whole animals and organs): s = survival time; bt = brain tumors; vt = various tumors; mt = mammary tumors; c = various chemical compounds; st = skin tumors; m = melatonin; h = hematology; ho = hormones; l = lymphoma.

[b] (+) = OK for parametric comparison, not sufficient for survival analysis.

References

1 ICNIRP (1998) Guidelines for limiting exposure to time-varying electric, magnetic, and electromagnetic fields (up to 300 GHz), *Health Physics*, **74**, 494–522.
2 Hansen, V.W., Bitz, A.K. and Streckert, J. (1999) RF exposure of biological systems in radial waveguides *IEEE Transactions on Electromagnetic Capability*, **41**, 487–493.
3 Ebert, S., Eom, S. J., Schuderer, J., Apostel, U., Tillmann, T., Dasenbrock, C. and Kuster, N. (2005) Response, thermal regulatory threshold and thermal breakdown threshold of restrained RF-exposed mice at 905 MHz, *Physics in Medicine and Biology*, **50**, 5203–5215.
4 Trosic, I., Busljeta, I., Kasuba, V. and Rozgaj, R. (2002) Micronucleus induction after whole-body microwave irradiation of rats, *Mutation Research*, **521**, 73–79.
5 Sommer, A.M., Bitz, A.K., Streckert, J., Hansen, V.W. and Lerchl, A. (2007) Lymphoma development in mice chronically exposed to UMTS-modulated radiofrequency electromagnetic fields, *Radiation Research*, **168**, 72–80.
6 Heikkinen, P., Kosma, V.M., Alhonen, L., Huuskonen, H., Komulainen, H., Kumlin, T., Laitinen, J.T., Lang, S., Puranen, L. and Juutilainen, J. (2003) Effects of mobile phone radiation on UV-induced skin tumourigenesis in ornithine decarboxylase transgenic and non-transgenic mice, *International Journal of Radiation Biology*, **79**, 221–233.
7 Repacholi, M. H., Basten, A., Gebski, V., Noonan, D., Finnie, J. and Harris, A.W. (1997) Lymphomas in E mu-Pim1 transgenic mice exposed to pulsed 900 MHZ electromagnetic fields, *Radiation Research*, **147**, 631–640.
8 Utteridge, T.D., Gebski, V., Finnie, J.W., Vernon-Roberts, B. and Kuchel, T.R. (2002) Long-term exposure of E-mu-Pim1 transgenic mice to 898.4 MHz microwaves does not increase lymphoma incidence, *Radiation Research*, **158**, 357–364.
9 Szmigielski, S., Szudzinski, A., Pietraszek, A., Bielec, M., Janiak, M. and Wrembel, J.K. (1982) Accelerated development of spontaneous and benzopyrene-induced skin cancer in mice exposed to 2450-MHz microwave radiation, *Bioelectromagnetics*, **3**, 179–191.
10 Szudzinski, A., Pietraszek, A., Janiak, M., Wrembel, J., Kalczak, M. and Szmigielski, S. (1982) Acceleration of the development of benzopyrene-induced skin cancer in mice by microwave radiation, *Archives of Dermatological Research*, **274**, 303–312.
11 Santini, R., Hosni, M., Deschaux, P. and Pacheco, H. (1988) B16 melanoma development in black mice exposed to low-level microwave radiation, *Bioelectromagnetics*, **9**, 105–107.
12 Heikkinen, P., Kosma, V.M., Hongisto, T., Huuskonen, H., Hyysalo, P., Komulainen, H., Kumlin, T., Lahtinen, T., Lang, S., Puranen, L. and Juutilainen, J. (2001) Effects of mobile phone radiation on X-ray-induced tumorigenesis in mice, *Radiation Research*, **156**, 775–785.
13 Imaida, K., Kuzutani, K., Wang, J., Fujiwara, O., Ogiso, T., Kato, K. and Shirai, T. (2001) Lack of promotion of 7,12-dimethylbenz[a]anthracene-initiated mouse skin carcinogenesis by 1.5 GHz electromagnetic near fields, *Carcinogenesis*, **22**, 1837–1841.
14 Huang, T.Q., Lee, J.S., Kim, T.H., Pack, J.K., Jang, J.J. and Seo, J.S. (2005) Effect of radiofrequency radiation exposure on mouse skin tumorigenesis initiated by 7,12-dimethybenz[alpha]anthracene, *International Journal of Radiation Biology*, **81**, 861–867.
15 Anane, R., Dulou, P.E., Taxile, M., Geffard, M., Crespeau, F.L. and Veyret, B. (2003) Effects of GSM-900 microwaves on DMBA-induced mammary gland tumors in female Sprague-Dawley rats, *Radiation Research*, **160**, 492–497.

16 Chou, C.K., Guy, A.W., Kunz, L.L., Johnson, R.B., Crowley, J.J. and Krupp, J.H. (1992) Long-term, low-level microwave irradiation of rats, *Bioelectromagnetics*, **13**, 469–496.

17 Wu, R.Y., Chiang, H., Shao, B.J., Li, N.G. and Fu, Y.D. (1994) Effects of 2.45-GHz microwave radiation and phorbol ester 12-O-tetradecanoylphorbol-13-acetate on dimethylhydrazine-induced colon cancer in mice, *Bioelectromagnetics*, **15**, 531–538.

18 Toler, J.C., Shelton, W.W., Frei, M.R., Merritt, J.H. and Stedham, M.A. (1997) Long-term, low-level exposure of mice prone to mammary tumors to 435 MHz radiofrequency radiation, *Radiation Research*, **148**, 227–234.

19 Frei, M.R., Berger, R.E., Dusch, S.J., Guel, V., Jauchem, J.R., Merritt, J.H. and Stedham, M.A. (1998) Chronic exposure of cancer-prone mice to low-level 2450 MHz radiofrequency radiation, *Bioelectromagnetics*, **19**, 20–31.

20 Frei, M.R., Jauchem, J.R., Dusch, S.J., Merritt, J.H., Berger, R.E. and Stedham, M.A. (1998) Chronic, low-level (1.0 W/kg) exposure of mice prone to mammary cancer to 2450 MHz microwaves, *Radiation Research*, **150**, 568–576.

21 Adey, W.R., Byus, C.V., Cain, C.D., Higgins, R.J., Jones, R.A., Kean, C.J., Kuster, N., MacMurray, A., Stagg, R.B., Zimmerman, G., Phillips, J.L. and Haggren, W. (1999) Spontaneous and nitrosourea-induced primary tumors of the central nervous system in Fischer 344 rats chronically exposed to 836 MHz modulated microwaves, *Radiation Research*, **152**, 293–302.

22 Chagnaud, J.L., Moreau, J.M. and Veyret, B. (1999) No effect of short-term exposure to GSM-modulated low-power microwaves on benzo[a]pyrene-induced tumors in rat, *International Journal of Radiation Biology*, **75**, 1251–1256.

23 Higashikubo, R., Culbreth, V.O., Spitz, D.R., LaRegina, M.C., Pickard, W.F., Straube, W.L., Moros, E.G. and Roti, J.L. (1999) Radiofrequency electromagnetic fields have no effect on the *in vivo* proliferation of the 9L brain tumor, *Radiation Research*, **152**, 665–671.

24 Adey, W.R., Byus, C.V., Cain, C.D., Higgins, R.J., Jones, R.A., Kean, C.J., Kuster, N., MacMurray, A., Stagg, R.B. and Zimmerman, G. (2000) Spontaneous and nitrosourea-induced primary tumors of the central nervous system in Fischer 344 rats exposed to frequency-modulated microwave fields, *Cancer Research*, **60**, 1857–1863.

25 Jauchem, J.R., Ryan, K.L., Frei, M.R., Dusch, S.J., Lehnert, H.M. and Kovatch, R.M. (2001) Repeated exposure of C3H/HeJ mice to ultra-wideband electromagnetic pulses: lack of effects on mammary tumors, *Radiation Research*, **155**, 369–377.

26 Zook, B.C. and Simmens, S.J. (2001) The effects of 860 MHz radiofrequency radiation on the induction or promotion of brain tumors and other neoplasms in rats, *Radiation Research*, **155**, 572–583.

27 Bartsch, H., Bartsch, C., Seebald, E., Deerberg, F., Dietz, K., Vollrath, L. and Mecke, D. (2002) Chronic exposure to a GSM-like signal (mobile phone) does not stimulate the development of DMBA-induced mammary tumors in rats: results of three consecutive studies, *Radiation Research*, **157**, 183–190.

28 La Regina, M., Moros, E.G., Pickard, W.F., Straube, W.L., Baty, J. and Roti, J.L. (2003) The effect of chronic exposure to 835.62 MHz FDMA or 847.74 MHz CDMA radiofrequency radiation on the incidence of spontaneous tumors in rats, *Radiation Research*, **160**, 143–151.

29 Anderson, L.E., Sheen, D.M., Wilson, B.W., Grumbein, S.L., Creim, J.A. and Sasser, L.B. (2004) Two-year chronic bioassay study of rats exposed to a 1.6 GHz radiofrequency signal, *Radiation Research*, **162**, 201–210.

30 Sommer, A.M., Streckert, J., Bitz, A.K., Hansen, V.W. and Lerchl, A. (2004) No effects of GSM-modulated 900 MHz electromagnetic fields on survival rate and

spontaneous development of lymphoma in female AKR/J mice, *BMC Cancer*, **4**, 77.

31 Yu, D., Shen, Y., Kuster, N., Fu, Y. and Chiang, H. (2006) Effects of 900 MHz GSM wireless communication signals on DMBA-induced mammary tumors in rats, *Radiation Research*, **165**, 174–180.

32 Tillmann, T., Ernst, H., Ebert, S., Kuster, N., Behnke, W., Rittinghausen, S. and Dasenbrock, C. (2007) Carcinogenicity study of GSM and DCS wireless communication signals in B6C3F1 mice, *Bioelectromagnetics*, **28**, 173–187.

6
Epidemiology
Joachim Schüz

6.1
Introduction

Epidemiology is concerned with the study of the occurrence and distribution of diseases in populations. Its ultimate goal is to learn about the causes of disease that may lead to effective preventive measures [1]. However, in contrast to experimental studies or clinical trials, epidemiological studies are usually observational, and are therefore vulnerable to bias and confounding. Thus, criteria are needed to assess whether observed empirical exposure–disease associations are possibly causal or more likely a play of chance or methodological artifacts [2]. Making sense of results from epidemiological studies is particularly challenging when they are conflicting or when there is a discrepancy between epidemiological and experimental findings.

A prominent example is the ongoing challenge of interpreting findings from epidemiological studies on exposure to extremely low-frequency (ELF) electromagnetic fields (EMFs) and the risk of childhood leukemia [3]. While the finding of a doubling of the leukemia risk has been observed in a number of epidemiological studies, little supportive evidence emerged from experimental research. The interpretation of the ELF EMF-related findings is particularly delicate because possible explanations range from methodological artifacts to a causal link and neither of these two contradictory explanations can be completely ruled out so far.

In this chapter, after an introduction to epidemiological study types and risk estimation, epidemiology's criteria for making sense of conflicting results are defined and exemplified using the ELF EMF and childhood leukemia studies.

6.2
Study Types and Risk Estimation

In principle, two basic epidemiologic approaches are intervention studies and observational studies. In the research field of hazard identification the vast majority are observational studies. The observational study types range from rather simple descriptive studies (e.g. ecological studies) to analytical studies like cross-sectional studies, case-control studies and cohort studies.

A basic descriptive measure in epidemiology is the incidence rate, which is calculated by dividing the number of new cases arising in a defined study population over a given time period by the total person-time at risk in that study population over that period; if death is the outcome of interest, mortality rates are calculated accordingly [1]. As the risk of many chronic diseases is strongly determined by age, an appropriate comparison of incidence rates needs to be made through age-standardized incidence rates; they take into account differences in the age distributions of the populations being compared. For similar reasons, incidence rates are usually given separately for males and females. Comparing age-standardized incidence rates of two populations (either defined by time or region or by exposure) can be done by calculating the ratio of the two rates or their difference (Figure 6.1; depending on whether the effect is multiplicative or additive); this results in an estimate of the relative risk or respectively the risk difference. A relative risk of 1 indicates that the effect under study is equally likely in both groups (also called no effect, null effect or negative association), whereas a relative risk greater than 1 indicates a risk and a relative risk less than 1 indicates a reduced risk or protective effect. If the relative risk is 1, the risk difference is zero. While the relative risk expresses the risk of an individual under exposure in comparison to the nonexposed situation, the risk difference displays rather a picture of the effect in the population. As shown in Figure 6.1, disease 1 is associated with a stronger risk increase for the individual (4-fold), while the

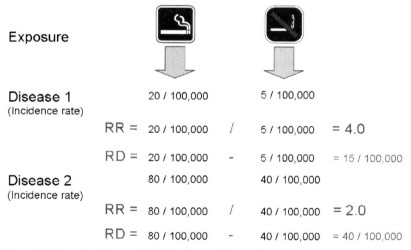

Figure 6.1 Hypothetical example of calculating relative risk (RR) and risk difference (RD) based on incidence rates of two diseases associated with smoking; smoking is associated with a stronger risk for the individual of disease 1, but more additional cases are due to smoking for disease 2.

impact on a population level due to exposure is larger for disease 2. In case-control studies the relative risk is approximated through the odds ratio, which is defined as the ratio of the odds of the outcome under study occurring in the exposed group to the odds of it occurring in the nonexposed group (from a 2×2 table).

In general, risk estimates derived from observational studies have to be interpreted with caution. In addition to that, in an overall risk assessment, the different observational study types contribute with different weights; more confidence relies on results derived from well-conducted prospective cohort studies than on results from case-control studies, whereas firm conclusions are rarely drawn from cross-sectional studies and particularly ecological studies. To illustrate this, the different types of studies – ecological, cross-sectional, case-control, cohort – will be briefly outlined together with their strengths and weaknesses.

In ecological studies exposure information is only available on an aggregate level, i.e. disease rates in populations are compared with each other, while one particular exposure level is assigned to each population irrespective of the exposure distribution within this population. A number of ecological studies were conducted on the association between radon and leukemia risk, comparing incidence or mortality rates across regions with differing average radon levels; however, the positive evidence from ecological studies received little support from recent improved approaches [4]. Ecological studies are only considered to be hypothesis generating and, hence, contribute little evidence in risk assessment. Findings of ecological studies need to be confirmed in analytical studies with a defined study hypothesis [1]. On the other hand, ecological studies may be informative when findings of analytical studies suggest a substantial effect, i.e. due to the widespread use of mobile phones in the population, a strong risk related to only moderate phone use would already have a measurable impact on time trends in incidence rates [5]. With regard to mobile phone use and brain tumor risk it may also be useful to analyze such time trends in subgroups since in the early 1990s the majority of mobile phone users were middle-aged males. As most of the microwaves emitted from a mobile phone are absorbed in the temporal lobes of the brain and as most mobile phone users prefer to use the phone on the right side of the head, an ecological analysis of the distributions of brain tumor localizations over time may also confirm a substantial risk, should it exist.

In a cross-sectional study, information on risk factors and on the effects of interest is obtained simultaneously. Cross-sectional studies are useful to estimate the prevalence of health conditions, but are not appropriate for studying chronic diseases. For example, in a cross-sectional study on cancer those cases with long survival would be over-represented, potentially introducing bias into the study [1].

Case-control studies are observational studies in which one group consists of persons having the disease of interest ("cases") and a suitable comparison group consists of persons free of this disease ("controls") (Figure 6.2). Cases and controls are then compared with regard to differences in their past exposures [1]. The nature of a case-control study is therefore retrospective as the direction of inquiry is from disease to exposure (reverse direction of causation). Advantages of case-control studies are that they are efficient in time and cost, and allow the investigation of

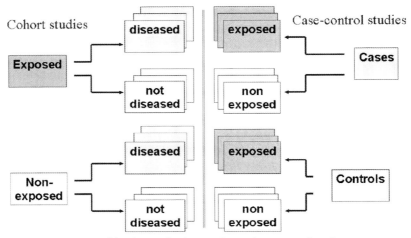

Figure 6.2 Overview of the design of cohort studies and case-control studies.

a wide range of possible risk factors of the disease of interest. They are often used for rare diseases. One challenge of case-control studies is to find a suitable control group which has to be representative of the total population from which the cases arose. Even when in some countries complete population registers provide an excellent frame for drawing a random sample of controls, most case-control studies require personal contact with study participants for assessing exposure and the representative nature is impaired by nonparticipants (participation bias or selection bias). In the childhood leukemia studies involving comprehensive ELF EMF measurements participation rates were usually in the range between 40 and 65% [6]. Another challenge is to obtain an accurate measurement of past exposures. Often, exposure is assessed during an interview or through a questionnaire; hence, the accuracy depends on the individual's ability to recall past events. A severe bias may be introduced when the reporting accuracy is systematically different between cases and controls (recall bias). However, objective exposure measurements may also lead to exposure misclassification, i.e. classifying exposed subjects as nonexposed and *vice versa* as they are only proxies of past exposures.

Cohort studies are regarded to be the superior study type as the direction of inquiry follows the nature of etiology, i.e. from cause to effect (Figure 6.2). In cohort studies groups are selected on the basis of their exposure status and then followed up in time with respect to occurrence of the diseases of interest. Hence, multiple outcomes can be studied for any one exposure [1]. Cohort studies are, however, very expensive and time consuming, especially for rare diseases with long induction periods. Challenges are how to update exposure status over time and how to avoid losses of follow up to protect the study against selection bias. Many cohort studies are retrospective cohort studies based on the use of pre-existing records for exposure assessment. The direction of inquiry in retrospective cohort studies is still prospective, but the date of entry into the cohort is in the past. Sometimes a cohort of exposed subjects is identified and the disease-specific incidence rates of the cohort are compared to the respective rates of the general population – a limitation of such studies is a possible healthy cohort effect. Schüz et al. [7] describe such an effect in a study on mobile phone subscribers followed up for cancer. Since the cohort of mobile phone subscribers had, on average, higher incomes and therefore comprised fewer smokers, the risk of lung cancer in the males of the cohort was observed to be almost 40% lower than in the general population; thus, the healthy cohort produced a spurious association between mobile phones and a reduced risk of lung cancer.

A general concern for all epidemiological study types is chance and confounding. The likelihood of chance associations increases with the increasing number of performed statistical tests or calculated confidence levels. Formal methods to adjust for multiple testing exist; however, their use increases the chance of missing a true association, which is why such adjustments are rarely seen in the epidemiological literature [8]. Since epidemiological studies are often costly, researchers should not be discouraged in using the material for explorative investigations, but the resulting risk estimates should be unambiguously earmarked as hypothesis generating. Confounding occurs when the association between exposure and outcome is mixed up with a real effect of another exposure on the same outcome, the two exposures being correlated [1]. It is possible to correct the relative risk estimates for the effects of confounding, given that accurate information on the confounding factor is available. Leuraud et al. [9] observed an increased lung cancer risk among uranium miners that was attenuated after adjustment for smoking; the study provides both evidence of smoking acting as a confounder in this relationship, but also that the association persists after taking the effect of smoking into account. Confounding by an unknown factor is a concern for every study; however, it needs to be discussed how likely this is.

6.3
Making Sense of Conflicting Results

When epidemiological data are used in risk assessment, criteria to evaluate the overall evidence are needed. The most widely accepted criteria are "Hill's" criteria, described in 1965 [10]. There are plenty of examples with suggestions of slightly modified versions [2], but the nature of the assessment suggested by Hill remains unchanged. In the following, the criteria temporality, strength, dose–response, consistency, specificity, absence of bias and confounding and biological plausibility will be discussed (Table 6.1).

6.3.1
Temporal Relation Consistent with Cause and Effect

It is beyond all question that exposure must precede the effect to be considered as a cause of it. A more difficult question is how much time has to elapse between cause

Table 6.1 Epidemiology's causal criteria.

Hill's criteria[a]	Criteria as discussed in this article[b]
1. Consistency (of association)	(Internal and external) Consistency (4)[c]
2. Strength (of association)	Strength of association (2)
3. Dose response	Dose response (3)
4. Temporality	Temporality (1)
5. Experimentation	
6. Specificity	Specificity (5)
7. Biologic plausibility	Biological plausibility (7) (including the discussion of coherence, experimentation, analogy)
8. Coherence	
9. Analogy	
	Absence of bias and confounding (6)

[a] Taken from Weed [2] as a listing of the original suggestions by Hill [10].
[b] Slightly modified criteria to better fit the example used for the illustration.
[c] Displays order in text.

and effect before an excess of this effect is observable in a study. When the effect is a chronic disease, one usually expects years or decades to pass before changes in the disease rates become apparent. Temporality is closely linked to biological plausibility as the biological pathway suggests the time period needed for the development of the disease. In the cohort study of the Japanese atomic bomb survivors, an excess of leukemias was observed years before excesses of solid tumors, indicating a shorter induction period after radiation exposure for leukemia [11]. In studies on mobile phones and the risk of brain tumors it was argued that both short- and long-term exposures may be of interest; if microwaves act as a promoter or progressor, an effect may become apparent already after few years of exposure [12]. Cohort studies follow the natural cause–effect path and are therefore particularly suitable for addressing temporality. In case-control studies, in which a reconstruction of the exposure history has to be made, it may sometimes be difficult to define the exact timing for the onset of exposure, e.g. because interviewees have difficulties in recalling the time period of first exposure.

In case-control studies on ELF EMFs and childhood leukemia, two main methods of exposure assessment were preferred: a calculation model for ELF EMFs based on historic power load information of power lines and contemporary measurements of ELF EMFs [6]. Historical field calculations allow the estimation of past exposures, in line with temporality, but they are feasible only for one ELF EMF source, i.e. high-voltage power lines. Measurements provide an adequate current picture of contributions from all ELF EMF sources, but as they are performed after the date of diagnosis (hence, after the effect occurred), it needs to be assessed how well they represent past exposures. It is likely that they reflect a point in time close to diagnosis better than exposure in infancy; hence, they are more in line with a temporal relationship of a short latency.

6.3.2
Strength of the Association

This is a quantitative criterion suggesting that a strong relationship between exposure and effect is more likely to be a causal effect. A meta-analysis is a useful tool for obtaining a combined risk estimate when there are several studies on the same topic [13]. However, if there is only a weak causal association, a lack of a strong empirical association cannot be used as a counter-argument against causality. The point estimates have to be interpreted in the context of the level of their confidence; a strong association with a high level of uncertainty may be less convincing than a more moderate association with narrow confidence boundaries. In line with this it was suggested to evaluate both strength of the association and level of significance [14].

The combined relative risk estimate from pooling the ELF EMF studies on childhood leukemia yielded a doubling in risk at average exposures of $0.4\,\mu T$ or higher [6], indicating a rather moderate association. Another pooled analysis found a similar risk increase, with a risk estimate of 1.7 at ELF EMF exposures above $0.3\,\mu T$ [15]. The two meta-analyses differed in a way that Greenland's approach [15] was to include every available study (hence, various methods of exposure assessment had to be combined into a single metric), while Ahlbom [6] pooled only studies with a defined population base using calculated ELF EMFs or long-term ELF EMF measurements for exposure assessment.

6.3.3
Dose–Response Relationship

Further evidence in favor of causality is provided if increasing levels of exposure are associated with an increasing strength of the effect. Unfortunately, many researchers only make use of testing for linearity, although statistical approaches are available and well described for allowing for a variety of possible shapes [16]. The issue of dose–response relationships has to be interpreted in the context of biological plausibility, whether one expects a linear relationship, a threshold effect, a U-shaped relationship or other even more complex scenarios. The relationship between low-dose ionizing radiation and cancer risk is an example for which several possible extrapolations are discussed, ranging from hormetic curves, thresholds to downwardly or upwardly curving increasing slopes, although linearity is suggested to be the most appropriate candidate [17].

The meta-analysis of ELF EMF and childhood leukemia studies of Ahlbom *et al.* [6] rather showed a threshold effect, with a doubling in risk at field levels of $0.4\,\mu T$ or higher, but no risk increase at lower field levels. A recent update of this meta-analysis [18] focusing on ELF EMF exposures at night showed a dose–response relationship that was actually indicative of linearity. As the same trend was seen using the exposure metric already applied in the first meta-analysis [6], it is likely that this trend is not due to the restriction of the exposure period to the night-time but to the exclusion of the studies using calculated ELF EMFs as exposure metric (in which no distinction between day and night was made) and the inclusion of an additional study from Germany [19]. However, data were too sparse to reliably predict the course of a dose–response curve for magnetic fields higher than 0.4 µT and the observed course was more or less compatible with trends ranging from a further increase in risk to a constant risk or a downward gradient at higher levels.

6.3.4
Consistency Within and Across Studies

If consistent results have been found across different studies, the association is more likely to be causal. There is, however, no clear-cut definition of how consistent study results have to be so they are regarded as such. From a statistical point of view, tests for heterogeneity are available, but it may be more appropriate to approach consistency from a qualitative perspective. Consistency of results could also be due to studies suffering from the same type and magnitude of bias; in this context it needs to be noted that a meta-analysis of such studies does not remove but promotes the bias [13]. Lack of consistency is not necessarily a sign of absence of causality, but may be due to biased studies showing no effect while the more sound studies show an effect. Examples of lack of internal inconsistency are risk estimates driven by implausible outliers or a strong dependency on the choice of the categorization of the exposure.

The consistency of the studies used in the meta-analysis of studies on ELF EMFs and childhood leukemia provides some evidence of a causal effect [6], and was a major reason for the classification of ELF EMFs as a possible carcinogen [20]. Additionally, the finding that grouping studies by certain study features with different strengths and limitations leads to similar results across these groups [6] has often been used as an argument in favor of consistency. In particular, the comparison of the studies conducted in the Nordic countries which do not include other ELF EMF sources than power lines, but have no participation bias, with the studies using ELF EMF measurements which use a more comprehensive exposure assessment method, but participation bias is a strong issue (see below), has been used to argue that selection bias may not fully explain the observed overall association [21]. However, most of the studies include only small numbers of exposed children, and the results are driven to some extent by the large US study [22], hence the question of consistency is debatable (Table 6.2).

6.3.5
Specificity

Specificity of an association is noticed when a particular exposure increases the risk of one disease, but not others. This has been used to strengthen the case in favor of causality, particularly in the evaluation of potential reporting bias in interview-based case-control studies. While a similar effect seen across all case groups would indicate reporting bias, an association specific to one case group provides some evidence in favor of causality, if appropriately balanced against the play of chance due to multiple statistical comparisons. A rational for Kaatsch *et al.* [23]

Table 6.2 Consistency across studies on ELF EMFs and risk of childhood leukemia: results from nine studies included in a pooled analysis of studies on residential magnetic fields and the risk of childhood leukemia (adapted from Ahlbom et al. [6]).

	Relative risk estimate with 95% confidence intervals ≥ 0.4 versus $<0.1\,\mu T$	Leukemia cases	
		Observed $\geq 0.4\,\mu T$	Expected[a] $\geq 0.4\,\mu T$
Canada	1.6 (0.7–3.7)	13	10
USA	3.4 (1.2–9.6)	17	5
UK	1.0 (0.3–3.4)	4	4
Norway	0 cases, 10 controls	0	3
Germany[b]	3.5 (1.0–12.3)	7	2
Sweden	3.7 (1.2–11.4)	5	2
Finland	6.2 (0.7–56.9)	1	0
Denmark	2 cases, 0 controls	2	0
New Zealand	0 cases, 0 controls	0	0
Total[b]	2.0 (1.3–3.2)	49	26

[a]Rounded to whole numbers.
[b]Update of Ref. [6] with a second German study [19].

to include childhood solid tumors as one additional control group for leukemia cases was to investigate which associations are specific to leukemia.

In respect of specificity, Ahlbom et al. [6] examined whether the observed association between ELF EMFs and childhood leukemia was stronger for acute lymphoblastic leukemia than for acute nonlymphoblastic leukemia. However, numbers for the latter group were too small to statistically reveal any differences.

6.3.6
Absence of Bias and Confounding

Noncausal explanations for observed exposure–effect associations include bias and confounding (described earlier). The demonstration of absence of substantial bias and confounding strengthens the evidence in favor of a true effect. However, bias and confounding may also mask a true effect and their impact has to be thoroughly discussed also for studies showing no associations.

With respect to confounding, various candidates have been proposed to explain the observed association between ELF EMFs and the risk of childhood leukemia, but none have been identified as true confounders to date [3]. Langholz [24] concluded from simulation studies that a relevant confounder in this context must be a relatively strong risk factor for childhood leukemia; other than rare genetic syndromes, no such risk factor has been established so far [25]. While confounding by an unknown factor is always a possibility taking the currently available knowledge into consideration, this seems to be an unlikely explanation.

With regard to exposure assessment methods there is a huge potential for exposure misclassification, i.e. classifying children as unexposed when they were in reality exposed or *vice versa*. Even long-term ELF EMF measurements in case-control studies only extended over 2 days at the most and averages of such measurements were used to categorize the entire past exposure of children. However, it is unlikely that the magnitude of misclassification was associated with disease status (being case or control). In such situations the expected bias in the relative risk estimate is towards an underestimation of an association; hence, it is very unlikely that the exposure misclassification apparent in the epidemiological studies was producing a spurious positive association.

With regard to selection bias, the low participation rates in many ELF EMF studies were of concern, especially in studies using measurements for exposure assessment [21]. Data suggested that families with a lower social status were particularly under-represented among controls [3]. In a simulation analysis using a German case-control data set [19] selection bias explained about 66% of the risk increase observed in this study. A considerable impact of selection bias has also been demonstrated for the large US study [26].

Taking all these considerations together, the probability is that selection bias alone is not sufficient to explain the entire association. However, taking into account the small increased risk is based on relatively small numbers of exposed children, a combination of selection bias, confounding and chance cannot be ruled out as a plausible explanation for the observed association.

6.3.7
Biological Plausibility

Epidemiological studies provide an estimate of the empirical association between exposure and effect. Hence, knowledge of a plausible biological mechanism or evidence from experimental results coherent with the epidemiological findings strongly supports positive empirical associations. As a minimum requirement, the introduction of a report of an epidemiological study should clearly state the study hypothesis and how it was addressed. Lack of biological plausibility weakens the epidemiological evidence, especially when associations are not strong; however, it requires that a reasonable number of sound experimental studies have been conducted. Epidemiologists like to point out that they observed the association between smoking and lung cancer years before it was accepted to be biologically plausible [27].

Lack of support from experimental studies is also hampering the interpretation of the epidemiological finding of a doubling of childhood leukemia risk with average ELF EMF exposures of 0.4 µT or higher [20]. However, the development of childhood leukemia is a unique process [25] and it is not certain whether the animal models used in genotoxicity or cell proliferation studies were adequate to test the proposed complex origin of childhood leukemia. There is strong evidence that the majority of cases of acute lymphoblastic leukemia in children are a result of two genetic "hits" and it has been proposed that the conversion rate of the preleukemic clone to overt leukemia is rather low, i.e. about 1% [25]. It is unclear whether this process requires a second direct DNA damage or whether some activation of cell proliferation of otherwise resting cells leads to further damage. In the latter case, factors qualifying for this activation are not necessarily mutagenic.

Kheifets *et al.* [28] reviewed recent suggestions to alternative explanations, but it appears that none of them reached a level beyond hypothesis. The suggestion that the magnetic fields suppress the nocturnal production and release of the hormone melatonin, which is assumed to have oncostatic capabilities [29], was not supported by a recent meta-analysis on night-time exposure to ELF EMFs and childhood leukemia [18]. Another hypothesis claims that contact currents occur with greater probability in residences with higher magnetic fields and that those contact currents lead to higher bone marrow doses of induced currents rather than magnetic field exposures alone [30]. However, there is little evidence to support the role of contact currents in leukemogenesis and it would also appear that this indirect mechanism must be extremely strong to be detected with only a weakly related level of residential magnetic field exposure.

6.4 Conclusions

A review of epidemiologic literature by Weed [31] suggests that criteria as defined by Hill [10] or derivations thereof are widely used. However, since the criteria list is a mixture of quantitative and qualitative criteria and lacks clear-cut definitions in many criteria as well as weights for each criteria, it is prone to subjective judgment. Holman *et al.* [32] conducted an experiment in which a group of epidemiologists had to apply causal criteria to evaluate an example of a possible association between a chemical agent and disease, and observed not only a general lack of agreement but also that certain criteria played a dominating role. This led Poole [33] in an accompanying editorial to suggest that the "criteria" are rather values or viewpoints, which is also more in line with their original proposal by Hill [10].

The application of Hill's criteria to the example of ELF EMF exposure and the risk of childhood leukemia is summarized in Table 6.3. This is in agreement with the conclusions of the epidemiological evidence drawn by the International Agency of Research on Cancer, finding limited evidence of an association between ELF EMFs and childhood leukemia due to consistency across studies, but an alternative explanation involving bias, chance and confounding could not be ruled out with reasonable confidence [20]. The body of literature suggests that there is no material disagreement when applying the single criteria to this prominent example; however, in their overall assessments there are reviews tending to suggest that the association between ELF EMF exposure and childhood leukemia is likely to be a true effect, while others strongly argument in favor of a spurious epidemiological association. This again provides evidence that epidemiology's causal criteria are rather a useful tool in risk assessment, but should not be overused as deterministic rules.

In conclusion, making sense of conflicting epidemiological data is a difficult task. Widely acknowledged criteria exist that ought to be used to evaluate the body of

Table 6.3 Application of the criteria of causality to the case of ELF EMFs and risk of childhood leukemia.

Criteria	Application
Temporality (1)[a]	fulfilled
Strength of association (2)	small to moderate
Dose response (3)	equivocal; not clear whether threshold or linearity at intermediate levels (0.2–0.4 µT), sparse data at high levels (≥ 0.4 µT)
Consistency (4)	relatively consistent
Specificity (5)	not assessable
Absence of bias and confounding (6)	strong evidence of selection bias, no evidence of confounding, some evidence of exposure misclassification
Biological plausibility (7)	little support

[a] Order as in text.

evidence. The major advantage of these criteria is that they provide a useful checklist to help the reviewer to make his or her viewpoint transparent for others. This also enables readers to match different risk assessments. It may happen, however, that even after their application, the question of whether an association is causal remains open.

For further reading it is recommended to take a look at the guidelines of how to report findings from epidemiological studies as suggested by the STROBE consortium [34] and curious readers may want to read the book by Judea Pearl with an exposition of modern analysis of causation.

References

1 Dos Santos Silva, I. (1999) *Cancer Epidemiology: Principles and Methods*, International Agency for Research on Cancer, Lyon.
2 Weed, D.L. (2005) Weight of evidence: a review of concept and methods, *Risk Analysis*, **25**, 1545–1557.
3 Schüz, J. (2007) Implications on protection guidelines from epidemiologic studies on magnetic fields and the risk of childhood leukemia, *Health Physics*, **92**, 642–648.
4 Laurier, D., Valenty, M. and Tirmarche, M. (2001) Radon exposure and the risk of leukemia: A review of epidemiological studies, *Health Physics*, **81**, 272–288.
5 Klaeboe, L., Lönn, S., Scheie, D., Auvinen, A., Christensen, H.C., Feychting, M., Johansen, C., Salminen, T. and Tynes, T. (2005) Incidence of intracranial meningiomas in Denmark, Finland, Norway and Sweden, 1968–1997, *International Journal of Cancer*, **117**, 996–1001.
6 Ahlbom, A., Day, N., Feychting, M., Roman, E., Skinner, J., Dockerty, J., Linet, M., McBride, M., Michaelis, J., Olsen, J.H., Tynes, T. and Verkasalo, P.K. (2000) A pooled analysis of magnetic fields and childhood leukaemia, *British Journal of Cancer*, **83**, 692–698.

7 Schüz, J., Jacobsen, R., Olsen, J.H., Boice, J.D., McLaughlin, J.K. and Johansen, C. (2006) Cellular telephone use and cancer risk: update of a nationwide Danish cohort, *Journal of the National Cancer Institute*, **98**, 1707–1713.

8 Rothman, K.J. (1990) No adjustments are needed for multiple comparisons, *Epidemiology*, **1**, 43–46.

9 Leuraud, K., Billon, S., Bergot, D., Tirmarche, M., Caer, S., Quesne, B. and Laurier, D. (2007) Lung cancer risk associated to exposure to radon and smoking in a case-control study of French uranium miners, *Health Physics*, **92**, 371–378.

10 Hill, A.B. (1965) The environment and disease: association or causation? *Proceedings of the Royal Society of Medicine*, **58**, 295–300.

11 Pierce, D.A., Shimizu, Y., Preston, D.L., Vaeth, M. and Mabuchi, K. (1996) Studies of the mortality of atomic bomb survivors. Report 12.1. Cancer: 1950–1990, *Radiation Research*, **146**, 1–27.

12 Lahkola, A., Tokola, K. and Auvinen, A. (2006) Meta-analysis of mobile phone use and intracranial tumors, *Scandinavian Journal of Work and Environmental Health*, **32**, 171–177.

13 Blettner, M., Sauerbrei, W., Schlehofer, B., Scheuchenpflug, T. and Friedenreich, C. (1999) Traditional reviews, meta-analyses and pooled analyses in epidemiology, *International Journal of Epidemiology*, **28**, 1–9.

14 Proctor, D.M., Otani, J.M., Finley, B.L., Paustenbach, D.J., Bland, J.A., Speizer, N. and Sargent, E.V. (2002) Is hexavalent chromium carcinogenic via ingestion? A weight-of-evidence review, *Journal of Toxicology and Environmental Health A*, **65**, 701–746.

15 Greenland, S., Sheppard, A.R., Kaune, W.T., Poole, C. and Kelsh, M.A. (2000) A pooled analysis of magnetic fields, wire codes, and childhood leukemia. Childhood Leukemia EMF Study Group, *Epidemiology*, **11**, 624–634.

16 Greenland, S. (1995) Dose–response and trend analysis in epidemiology – alternatives to categorical analysis, *Epidemiology*, **6**, 356–365.

17 Brenner, D.J., Doll, R., Goodhead, D.T., Hall, E.J., Land, C.E., Little, J.B., Lubin, J.H., Preston, D.L., Preston, R.J., Puskin, J.S., Ron, E., Sachs, R.K., Samet, J.M., Setlow, R.B. and Zaider, M. (2003) Cancer risks attributable to low doses of ionizing radiation: assessing what we really know, *Proceedings of the National Academy of Sciences of the United States of America*, **100**, 13761–13766.

18 Schüz, J., Svendsen, A.L., Linet, M.S., McBride, M.L., Roman, E., Feychting, M., Kheifets, L., Lightfoot, T., Mezei, G., Simpson, J. and Ahlbom, A. (2007) Nighttime exposure to electromagnetic fields and childhood leukemia: an extended pooled analysis, *American Journal of Epidemiology*, **166**, 263–269.

19 Schüz, J., Grigat, J.P., Brinkmann, K. and Michaelis, J. (2001) Residential magnetic fields as a risk factor for childhood acute leukemia: results from a German population-based case-control study, *International Journal of Cancer*, **91**, 728–735.

20 IARC (2002) *IARC Monographs on the Evaluation of Carcinogenic Risks to Humans: Volume 80. Non-ionizing radiation, Part 1: Static and Extremely Low-frequency (ELF) Electric and Magnetic Fields*, International Agency for Research on Cancer, Lyon.

21 Mezei, G. and Kheifets, L. (2006) Selection bias and its implications for case-control studies: a case study of magnetic field exposure and childhood leukemia, *International Journal of Epidemiology*, **35**, 397–406.

22 Linet, M.S., Hatch, E.E., Kleinerman, R.A., Robison, L.L., Kaune, W.T., Friedman, D.R., Severson, R.K., Haines, C.M., Hartsock, C.T., Niwa, S., Wacholder, S. and Tarone, R.E. (1997) Residential exposure to magnetic fields and acute lymphoblastic

leukemia in children, *New England Journal of Medicine*, **337**, 1–7.

23 Kaatsch, P., Kaletsch, U., Meinert, R., Miesner, A., Hoisl, M., Schüz, J. and Michaelis, J. (1998) German case control study on childhood leukemia – basic considerations, methodology and summary of the results, *Klinische Pädiatrie*, **210**, 185–191.

24 Langholz, B. (2001) Factors that explain the power line configuration wiring code-childhood leukemia association: what would they look like? *Bioelectromagnetics Supplement*, **5**, S19–S31.

25 Greaves, M. (2006) Infection, immune responses and the aetiology of childhood leukemia, *Nature Reviews Cancer*, **6**, 193–203.

26 Hatch, E.E., Kleinerman, R.A., Linet, M.S., Tarone, R.E., Kaune, W.T., Auvinen, A., Baris, D., Robison, L.L. and Wacholder, S. (2000) Do confounding or selection factors of residential wiring codes and magnetic fields distort findings of electromagnetic fields studies? *Epidemiology*, **11**, 189–198.

27 Doll, R., Peto, R., Boreham, J. and Sutherland, I. (2004) Mortality in relation to smoking: 50 years' observations on male British doctors, *British Medical Journal*, **328**, 1519–1528.

28 Kheifets, L., Repacholi, M., Saunders, R. and van Deventer, E. (2005) The sensitivity of children to electromagnetic fields, *Pediatrics*, **116**, e303–e313.

29 Henshaw, D.L. and Reiter, R.J. (2005) Do magnetic fields cause increased risk of childhood leukemia via melatonin disruption? *Bioelectromagnetics Supplement*, **7**, S86–S97.

30 Kavet, R. (2005) Contact current hypothesis: summary of results to date, *Bioelectromagnetics Supplement*, **7**, S75–S85.

31 Weed, D.L. (1997) On the use of causal criteria, *International Journal of Epidemiology*, **26**, 1137–1141.

32 Holman, C.D.J., Arnold-Reed, D.E., de Klerk, N., McComb, C. and English, D.R. (2001) A psychometric experiment in causal inference to estimate evidential weights used by epidemiologists, *Epidemiology*, **12**, 246–255.

33 Poole, C. (2001) Causal values, *Epidemiology*, **12**, 139–141.

34 von Elm E., Altman D.G., Egger M., Pocock S.J., Gøtzsche P.C., Vandenbroucke J.P., STROBE Initiative (2007) The Strengthening the Reporting of Observational Studies in Epidemiology (STROBE) statement: guidelines for reporting observational studies. *Lancet*, **370**, 1453–1457.

35 Pearl J. (2000) Causality: Models, Reasoning, and Inference. Cambridge University Press.

7
Principles and Practice of Evidence Characterization in Environmental Clinical Case Studies

Caroline Herr and Thomas Eikmann

7.1
Clinical Environmental Medicine

Clinical environmental medicine centers on the assessment of environmental exposure and health impairment in individual patients. It was established about 1995 in Germany and other German-speaking countries. Presently, in Germany, general practitioners and physicians of different fields can participate in a postgraduate training in "environmental medicine" [1] comprising a defined theoretical training and internship.

The great number of possible environmental exposures and scenarios as well as the multitude of possible health effects determine the interdisciplinarity in training and the need for cooperation in case assessment in environmental medicine [2]. In evidence characterization of environmental cases physicians must be able to appreciate aspects of microbiology, hygiene, toxicology, technology and epidemiology, and from a clinical perspective nearly all fields, especially allergology, infectiology, occupational as well as general medicine, pediatrics and oncology. Figure 7.1 gives an overview of the relevant aspects concerning assessment in clinical cases in environmental medicine.

7.2
Assessment of Health Complaints

The process of evidence characterization begins with the patient's history. Systematic assessment of a case history is most efficiently performed via a physician interview based on a standardized questionnaire developed specifically for environmental medicine. This questionnaire must comprise the present health complaints and impairments as well as their development. A special focus should be laid on the initial situation in which the presently reported complaints first occurred and situations

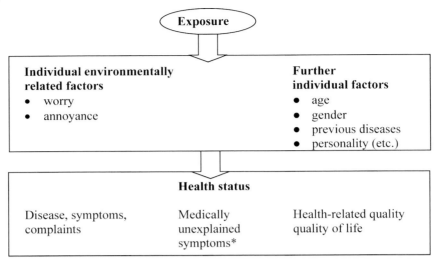

Figure 7.1 Overview of the relevant aspects concerning assessment in clinical cases in environmental medicine.
*Health symptoms for which a physician has so far not been able to find an explanation in clinical diagnostics.

leading to aggravation or amelioration, i.e. holidays. Health reports of others living in the same environment should be considered.

7.2.1
Environmental Attribution

During systematic questionnaire-based case assessment the environmental exposure (s) to which the patients attribute their complaints must be noted. Patients presenting with an environmentally related health condition will usually have definite and in part fixed ideas about the environmental factors causing their disease. This might give the physician valuable information concerning relevant exposures. The physician needs to take the patient's environmental attribution into consideration even if there is no evidence for a causal role of this environmental factor in health impairment. It will be necessary to consider the patient's environmental attribution in the process of the ongoing medical consultation in order to perform an adequate case-related risk communication [3]. If the physician neglects to assess and consider the patient's environmental attribution at this point the following diagnostics and evaluation will only consider environmental exposures resulting from the medical history (see below) and will not be related to the patient's individual concerns (Figure 7.1). If the environmentally related disease is eventually not confirmed in the course of diagnostics the patients will most likely not be inclined to accept the physicians evaluation if this evaluation did not consider environmental factors indicated by the patient.

7.2.2
Case History

A detailed history of physicians' previous findings and medical procedures may give hints for underlying medical conditions appropriate to fully or in part explain the present health complaints. Individual susceptibility, i.e. atopic disposition, will also become evident. Furthermore, possible relevant environmental exposures due to housing, workplace, leisure time activities, nutritional habits and personal care or cleaning products derived from the questionnaire-based interview and possible previously performed environmental diagnostics, i.e. measurements of exposure, must be taken into consideration. The history taking must be finalized with a detailed physical inspection.

At this point the physician will also be able to judge in which manner the patient's environmental attribution is fixed and supported or based on information given by previously consulted physicians or other information sources.

The physician will now need to evaluate, in an evidence-based approach, whether the environmental exposure suspected by the patient and/or derived from the case history and/or previous findings is adequate in character and timing to possibly explain in total or part the health complaints or impairment stated by the patient.

7.3
Exposure Assessment and Evaluation

In cases in which the patient's report and the case history give reason for the physician to consider exposure analyses, several necessary procedures alone or in combination are conceivable.

7.3.1
Biomonitoring

Biomonitoring aims at determining the internal body dose of a substance or its metabolites [4]. Usually it is performed in whole blood, serum, or spot or 24-h urine. Respective laboratories should participate in quality assurance programs [5]. Materials other than those aforementioned are not considered useful for case diagnostics. The results of these measurements need to be classified as to their toxicological relevance. In Germany, the Federal Environmental Agency regularly publishes threshold values (Humanbiomonitoring Werte [6]) for selected substances. Substances for which no toxicological thresholds have been published by national or international committees may make it necessary for the specialist to refer to the scientific databases and literature in order to classify an individual's exposure. This usually poses many uncertainties [7] and cannot be performed in the scope of a routine physician examination. In these cases specialized centers, i.e. state or university medical centers for environmental medicine, should be involved in the exposure evaluation or, better yet, in the planning and performance of the biomonitoring diagnostics.

7.3.2
Effect and Susceptibility Monitoring

As stated above, the clinical case assessment must consider individual susceptibility in the evaluation of an environmental exposure. An asthmatic patient with various inhalationary allergies will be at a greater risk for developing further allergies and consecutive health deterioration in an apartment with substantial (more than $0.5\ m^2$) indoor mold and dampness than a healthy person. So far, specific laboratory methods of susceptibility monitoring for everyday environmental exposures to chemicals have not been established. Although the idea of genetically determined environmental susceptibility is attractive, the Commission on Methods and Diagnostics in Environmental Medicine at the German Robert-Koch Institute has published a statement concerning, genetic susceptibility [8] and its inappropriateness in clinical case assessment and evidence characterization.

Effect monitoring aims to disclose "effects" of environmental exposures in the body by using different laboratory methods, e.g. changes in liver enzymes due to exposure to wood preservatives. Effects of substances in environmental concentration cannot usually be measured by routine laboratory measurements, as the exposures are too low, and the measured effect parameters are not specific and can be altered by various factors. Although effect monitoring is commonly used by some physicians in environmental medicine, most methods cannot be considered as state of the art in evidence-based medicine and are not appropriate for evidence characterization. This applies, for instance, for blood counts or melatonin measurements in case assessment of electrosensitivity [9, 10].

7.3.3
On-site Inspection

In cases where special scenarios, i.e. indoor environment of a private home, seems to cause or partly cause health problems the physician should consider an on-site inspection in order to get a realistic impression of possible environmental exposures in a patient's home environment and in the surrounding neighborhood. These inspections also give the physician the opportunity to study a patient in his private setting and gain further information on competing, nonenvironmental factors possibly impacting reported health status. It will often be useful to perform the on-site inspection together with specialists of other, technical fields, e.g., construction for appropriate case assessment of building-related factors.

7.3.4
Ambient Monitoring

During or following an on-site inspection ambient monitoring, e.g., analyses of building materials or indoor air, can be performed to search for possible sources of environmental contamination. Laboratories and institutions assigned with respective analyses should perform these according to published guidelines, e.g.

the Association of German Engineers (www.vdi.de/vdi/english). Analyses should be based on a matching health and environmental exposure history. For example, measurements of wood preservatives in indoor air are not useful if a patient complains of chronic back pain and the on-site inspection reveals wooden furnishing in a seldom used storage room in the basement of a house. In this case the complaint does not match the suspected substance and a relevant exposure is not likely, as the room with possible contamination is not in frequent use. If, in spite of these inconsistencies, indoor air measurements are performed and levels of wood preservatives, i.e. pentachlorophenol, are detected in the indoor air of the storage room, it will be difficult to explain to the patient that these substances are generally found in traces everywhere and cannot be made responsible for his/her health complaints.

7.4
Interdisciplinary Clinical Diagnostics

Usually clinical cases with environmental complaint attribution will present with a multitude of health complaints in different organ systems and a great number of clinical reports from various specialists [11]. Environmental assessment and risk characterization with specific history taking, and bio- and ambient monitoring will usually not address and explain the complaints of these clinical cases adequately and completely. Therefore, interdisciplinary clinical diagnostics should be performed to reveal all current clinical conditions of the patient and their relevance for the health status (see Box 7.1). In order to consider nonsomatic causes of health complaints, somatic clinical diagnostics should always be supplemented by psychosomatic exploration.

This type of procedure will enable the treating physician to consider all relevant medical aspects in case assessment. When communicating with the patient, it will then be possible to hint at possible alternative explanations for their health impairments, as less than 7% of clinical cases with environmental complaint attribution will eventually be confirmed to have complaints caused by environmental factors [11].

Box 7.1
Clinical patient management – Hessian Center for Environmental Medicine, University of Giessen, Germany [11]

Interdisciplinary clinical diagnostic starts by taking the environmental/toxicological history in the Environmental Outpatient Department of the Institute for Hygiene and Environmental Medicine. An environmental medical questionnaire and previous medical or environmental monitoring records are the basis for further clinical investigations in the University Medical Center, Giessen or further environmental diagnostics. The environmental attributions of the patients are also assessed.

The diagnostic protocol always includes a dermatological and two psychosomatic consultations during an optional in-patient stay in the Department of Dermatology. Additionally, patients are sent to specialists if unclear diagnostic results have been obtained previously or if appropriate referrals have not been considered. Subsequently, the medical findings of the patients are discussed in an interdisciplinary case conference with all participating physicians of the University Medical Center. At this point the patients' conditions are specified and ranked by their likelihood to explain the complaints. It is also laid down whether any of the previous or present findings point at an environmental cause of the complaints. Furthermore, advice for future treatment is given.

All previous and present medical conditions are classified according to the International Statistical Classification of Diseases and Related Health Problems 10th Revision (ICD-10) [12]. The patients are grouped according to whether physicians had found clinical conditions that could explain their chronic health complaints medically. They are classified as suffering from medically unexplained symptoms (if in the case conferences the clinical conditions found are not considered to explain their chronic symptoms). Psychosomatic conditions (F^* according to ICD-10) merely describing the symptoms are not considered explanatory for chronic complaints in this context.

In a final appointment the patients are informed about the findings. Here, environmental aspects of their illnesses are once again specifically addressed and it is clearly communicated whether there is scientifically based evidence for a toxicologically relevant environmental exposure.

7.4.1
Evaluation of Clinical Cases

As in other clinical fields the evaluation process must be completed with a written report on the physician's findings (case history, exposure assessment, clinical evaluation) and advice.

Clinical cases must be evaluated on the basis of observations and findings. As described above, evaluation can be derived from case history (individual susceptibility!) as well as biomonitoring, ambient monitoring and clinical findings with evidence-based diagnostic methods (see Boxes 7.2 and 7.3). In environmental medicine exposure avoidance is sometimes part of the advised therapy. The costs resulting from exposure avoidance are usually not paid for by the medicare system and can be quite high, e.g., changes in housing, making it all the more necessary that the advice is evidence based. The same principle applies for preventive measures advised. Both must be related to a realistic health risk from which an individual is to be protected. Lowering of exposure to high-frequency electromagnetic fields, i.e. from mobile stations, cannot be advised based on risk assessment from scientifically based studies, as laboratory animal and/or epidemiological studies have not shown the potential of these exposures to cause health deterioration in the concentration levels in question.

> **Box 7.2**
> **Case Report**
>
> A 44-year-old saleswoman – presently unfit for work – presented in the Environmental Outpatient Department of the Institute for Hygiene and Environmental Medicine, University of Giessen.
>
> *Health complaints*
> She reported having the following complaints within the past 5 years: altered temperature sensation with heat waves and reddening of the thorax and neckline as well as nightly sweating up to 6 months ago, deteriorating ability to perform, fatigue, muscle weakness, hoarseness and scratching of the throat, headache, periorbital swelling, wheals, sleep apnea and snoring. During the past 6 months she gained 6 kg in weight.
>
> *Environmental factors the patient attributed her complaints to*
> Plastics from packaging and formaldehyde at her workplace, electromagnetic fields, water vein and iron construction in housing.
>
> *Previous medical reports*
> She presented eight medical reports from specialists for internal medicine, gynecology, radiology, traditional Chinese medicine and oral surgery with 12 different diagnoses, e.g., primary sterility with now postmenopausal hormone status, hypercholesteremia, irritable stomach, surgery for an abscess near the upper jaw, etc.
> Her family physician certified that she suffered from "higher-grade electrosensitivity" and "intolerance of environmental chemicals". The physician had treated this with detoxification ("Ausleitung") and stated the therapy would only remain successful if the patient was not exposed to electromagnetic fields and environmental chemicals at work.
>
> *Previous environmental findings*
> Report of a measurement of water pipes in the patient's home.
>
> *Medication*
> Her present medication: progesterone and estrogenic hormone substitution and various homeopathic and herbal medication for eyes, intestine, liver, gall bladder, general performance, supplements: vitamin E and selenium.
>
> *Family history*
> She reported being separated from her husband for 7 years and living with a new partner, who suffered from multiple sclerosis.

Box 7.3
Case Report: Interdisciplinary Clinical Diagnostics in the Hessian Centre for Environmental Medicine

Diagnoses
Environmental medicine: no indication for an environmentally related disease or relevant environmental exposure from the interview and specific questionnaire.

Medical consultations
Dermatology, gynecology, neurology, ear, nose and throat, orthopedics, internal, and dental.

Diagnoses (as ranked in the clinical case conference; see Box 7.1)

1. Undifferentiated somatization
2. Intolerance to nickel(II) sulfate
3. Tendinosis calcarea right shoulder
4. Spondylarthosis cervico-thoracal
5. Caries and apical periodontitis
6. Early postmenopausal status
7. Uterus myomatosus

Advice (following the clinical case conference with the consulting specialists)

- In-patient psychosomatic therapy
- Avoid exposure to nickel(II) sulfate
- Possible surgery of the right shoulder in case of more intense pain
- Dental overhaul

Communication with the patient

- No indication for environmental causes of the health complaints.
- Electrosensitivity and multiple chemical sensitivity are not established diagnoses, and various studies have not been able to relate exposures in environmental levels to the multitude of symptoms presumed to be caused by these exposures.
- None of the reported health complaints can be explained by the somatic findings. Therefore, in addition to somatic therapy, psychosomatic therapy seems adequate to help the patient to cope with the multitude of complaints and perhaps eventually return to work.

7.5 Conclusions

In environmental medicine evidence characterization of the relationship between exposure and health impairment can often not be based on textbook knowledge or statements of national or international committees since the alleged exposures vary over location and time due to changes in the production process, technology and risk perception of the population. This makes it necessary for physicians in some cases to rely on "primary" literature, i.e. scientific studies and newly developed diagnostic methods, when evaluating a clinical environmental case. The general criteria for evaluation of scientific work, biological plausibility, reproducibility of findings by different working groups, published in peer-reviewed journals, etc., must be applied, when using "new" methods in evidence characterization and therapy of environmental cases. This can usually only be performed by specialists with a scientific interest in the field. For an appropriate risk characterization a physician not specialized in environmental medicine should consider referring cases with presumed environmentally related health impairment, especially if new techniques or substances are involved, to specialized centers.

References

1 Bundesärztekammer (2006) *Strukturierte curriculäre Fortbildung Curriculum "Umweltmedizin"*, Bundesärztekammer, Berlin.
2 Herr, C. and Eikmann, T. (1998) [Environmental medicine – interdisciplinary cooperation encouraged. Experience of the Hessen Center for Clinical Environmental medicine], *Fortschritte der Medizin*, **116** 16–18.
3 Eis, D. (2000) Methoden und Qualitätssicherung in der Umweltmedizin, *Bundesgesundheitsblatt – Gesundheitsforschung – Gesundheitsschutz*, **43**, 336–342.
4 Umweltbundesamt (2000) Referenz- und Human-Biomonitoring-(HBM)-Werte, *Umweltmedizinscher Informationsdienst*, **1**, 9–12.
5 G-EQUAS (2007) *The German External Quality Assurance Scheme*, G-EQUAS, Erlangen [http://www.g-equas.de] [Retrieved: 18.07.2007].
6 Umweltbundesamt (2007) *Extreme Values and other Criterias*, Umweltbundesamt Berlin [http://www. umweltbundesamt. de/gesundheit/monitor/definitionen. htm] [Retrieved: 16.08.2007].
7 TERA (2007) *International Toxicity Estimates for Risk Database*, Toxicology Excellence for Risk Assessment, Cincinnati, OH [http://www. Tera.org/iter/] [Retrieved: 16.08.2007].
8 Mitteilung der Kommission "Methoden und Qualitätssicherung in der Umweltmedizin" (2004) Genetische Polymorphismen (Sequenzvariationen) von Fremdstoff-metabolisierenden Enzymen und ihre Bedeutung in der Umweltmedizin, *Bundesgesundheitsblatt – Gesundheitsforschung – Gesundheitsschutz*, 47, 1115–1123.
9 Mitteilung der Kommission "Methoden und Qualitätssicherung in der Umweltmedizin" (2006) Parameter des roten Blutbildes bei Exposition durch Mobilfunkanlagen, *Bundesgesundheitsblatt – Gesundheitsforschung – Gesundheitsschutz*, 49, 833–835.
10 Mitteilung der Kommission "Methoden und Qualitätssicherung in der Umweltmedizin" (2005) Melatonin in der umweltmedizinischen Diagnostik im Zusammenhang mit elektromagnetischen Feldern, *Bundesgesundheitsblatt – Gesundheitsforschung – Gesundheitsschutz*, 48, 1406–1408.
11 Herr, C.E., Kopka, I., Mach, J., Runkel, B., Schill, W.B., Gieler, U. and Eikmann, T.F. (2004) Interdisciplinary diagnostics in environmental medicine findings and

follow-up in patients with chronic medically unexplained health complaints, *International Journal of Hygiene and Environmental Health*, **207**, 31–44.

12 WHO (2003) *International Classification of Diseases (ICD)*, World Health Organization, Geneva [http://www.who.int/whosis/icd10/] [Retrieved: 16.08.2007].

III
Making Sense of Conflicting Data: Procedures for Characterizing Evidence

8
Characterizing Evidence with Evidence-based Medicine
Alexander Katalinic and Dagmar Lühmann

8.1
What is Evidence-based Medicine?

The concept of "evidence-based medicine" (EbM) was introduced by David Sackett during the 1970s at McMaster University, Hamilton, Canada. Sackett and colleagues defined EbM as follows ([1], p. 71):

> Evidence-based medicine is the conscientious, explicit and judicious use of current best evidence in making decisions about the care of individual patients. The practice of evidence-based medicine means integrating individual clinical expertise with the best available external clinical evidence from systematic research.

How is this concept applied to actual patient care? It implies that expert physicians should base their clinical decisions on both their individual clinical expertise (experience) as well as on the best available external evidence (from evaluative clinical research). Neither one alone is enough. Clinical expertise is required to decide whether a piece of evidence is applicable in a specific decisive situation. Without this expertise, applying "evidence" may lead to inappropriate patient care. At the same time, solely relying on clinical expertise carries the risk of applying outdated care – to the detriment of patients [1] (Centre for Evidence Based Medicine; http://www.cebm.net).

This basic concept of evidence-based decision making was updated by Haynes *et al.* [2]. The new concept integrates:

(1) "Clinical state and circumstance" (individual expertise – internal evidence)
(2) "Research evidence" (external evidence)
(3) "Patients' preferences and actions"

It should be noted that the weight given to each of these components may vary according to the circumstances of a given clinical case. For example, even if strong

---

The Role of Evidence in Risk Characterization: Making Sense of Conflicting Data.
Edited by Peter M. Wiedemann and Holger Schütz
Copyright © 2008 WILEY-VCH Verlag GmbH & Co. KGaA, Weinheim
ISBN: 978-3-527-32048-6

external evidence suggests marked benefits for a specific indication (from the perspective of the physician) its application may be overruled by the patient's preferences (e.g. patient refusal of chemotherapy after nodal-positive breast cancer). In another situation there may be only methodologically weak external pieces of evidence available to support a decision (e.g. for the treatment of a rare disease). In this case clinical expertise may overrule (weak) research results.

These examples demonstrate that EbM should not be termed "cookbook medicine". "Evidence does not make decisions – people do" [2].

8.2
EbM Process

Basically EbM is a clinical approach [evidence-based clinical medicine (EbCM)], although its methodology nowadays is more commonly applied in healthcare decision making (evidence-based healthcare).

The starting point for EbCM is posed by a problem in clinical decision making. For example, on the occasion of his periodic health examination the following test results are found in an asymptomatic 55-year-old male patient:

- Total cholesterol 320 mg/dl (reference < 240)
- Blood pressure: 130/90 mmHg (reference 120/80)
- Body Mass Index: 31 (reference 20–24)
- The patient smokes about 20 cigarettes a day; his father died from a heart attack at the age of 58

At the follow-up visit he is told that with his combination of risk factors his probability of developing a manifest coronary heart disease (CHD) within the next 10 years amounts up to about 30%. In order to lower this risk the doctor advises him to take a statin drug.

The patient, who does not like the idea of being "on a drug", asks his doctor: "*What's your evidence* that this will do any good for me?".

8.3
Five Steps of EbM

EbM defines five steps to systematically answer such questions [3]:

(1) Transform the clinical problem into a precise and answerable question (asking answerable questions)
(2) Search the scientific literature for relevant research results (finding the best available evidence)
(3) Evaluate the evidence for its validity and usefulness (critical appraisal)
(4) Implement useful findings in clinical practice (acting on the evidence)
(5) Evaluate your performance

8.3.1
Asking Answerable Questions

This is an important step and the key to the following literature search. The precise question should reflect the clinical situation: Patients' characteristics, Intervention, Control situation and relevant Outcome parameters (PICO schema for question).

In our example the answerable question could be: "In an asymptomatic 55-year-old male with multiple risk factors for CHD, will treatment with a statin drug compared to no treatment lower the patient's risk of experiencing a possibly fatal heart attack?"

8.3.2
Finding the Best Available Evidence

Usually in a literature search the most relevant research results have to be identified. In this context "best available evidence" refers to the probability that the results reported in a piece of scientific literature are unbiased. The first safeguard against biased results is the choice of the adequate study design.

In the search for proof of a treatment's efficacy, an experimental study design, ideally a randomized controlled trial (RCT), will yield the most reliable results. In this type of study the results of two patient groups are compared: group A receives the intervention under investigation (active treatment) and group B receives a control intervention. Allocation of patients to the study groups is performed by randomization, in order to evenly distribute individuals with confounding factors (e.g. smokers) between the two groups. Distortion of results may furthermore be controlled by "blinding" trial participants (doctors, patients, outcome assessors and statisticians) against the knowledge of which group receives the active treatment. It is common understanding that the results of RCTs near enough prove a causal association between treatment and outcome.

If for the treatment in question no RCTs can be found, we would like to base our decision on results from the next best study design which, in the case of treatment, would be a prospective cohort study. If there are not any cohort studies either, we would use results from case-control studies. If these are missing as well, the decision will have to consider the results of case series or case reports. Finally, if there are no data from clinical research available, expert opinion or inferences from bench research will constitute the "best available" evidence.

Putting study designs into an order according to their susceptibility for biased results means establishing a "Hierarchy of Evidence" or "Levels of Evidence" (LoE). Table 8.1 shows the commonly used evidence levels according to the Oxford Centre of EbM (www.cebm.net). For each area of clinical decision making (determining etiology, assessing efficacy of therapy or preventive measures, establishing a prognosis, establishing a diagnosis) the supporting evidence is classified into five levels (with different sublevels), where level 1 stands for highest (i.e. lowest susceptibility for biased results), level 5 for lowest (i.e. highest

Table 8.1 Levels of evidence (adapted from the Oxford Centre for EbM).

Level of evidence		Therapy, prevention, etiology, harm	Prognosis
1	a	systematic review of RCTs	systematic review of inception cohort studies
	b	individual RCT with narrow confidence interval	individual inception cohort with more than 80% follow-up
	c	all or none – trial	all or none – case series
2	a	systematic review of cohort studies	systematic review of either retrospective cohort studies or untreated control groups in RCTs
	b	individual cohort study including low-quality RCT	retrospective cohort study or follow-up of untreated control patients in an RCT
	c	"outcomes" research; ecological studies	"outcomes" research
3	a	systematic review of case-control studies	
	b	individual case-control study	
4		case-series and poor quality cohort and case-control studies	case-series (and poor quality prognostic cohort studies)
5		expert opinion without explicit critical appraisal, or based on physiology, bench research or "first principles"	expert opinion without explicit critical appraisal, or based on physiology, bench research or "first principles"

Adapted from www.cebm.net, by Bob Phillips, Chris Ball, Dave Sackett, Doug Badenoch, Sharon Straus, Brian Haynes and Martin Dawes since November 1998; for the full list of evidence levels see: http://www.cebm.net/levels_of_evidence.asp.

susceptibility for biased results) evidence level. The best evidence (LoE 1a) to support an etiological or therapeutical decision would be the results of a systematic review of high-quality RCTs. LoE 1b is presented by a single high-quality RCT with narrow confidence intervals. Please note that in support of clinical problems other than therapeutic decisions different study design might be appropriate, e.g. "best evidence" to support a prognostic forecast would be results from prospective cohort studies.

For the statin example outlined above the literature search should ideally retrieve a systematic review of RCT results. This review will certainly contain the results of the randomized controlled West of Scotland Coronary Prevention Study Group (WOSCOPS) study [4]. These suggest that in comparison to placebo treatment, treatment with a statin drug significantly lowers the risk of dying from CHD in high-risk asymptomatic male study participants. The WOSCOPS study by itself would be classified as LoE 1b; if embedded into a systematic review, the classification would be LoE 1a according to the Oxford criteria.

However, retrieving a study with a high-quality design does not automatically ensure that the provided evidence is actually strong. The results of any trial can be seriously compromised by methodological deficiencies. Bias, confounding or other methodical issues may be present and distort a trial's results. Therefore it is essential to carefully check methodological study quality before accepting trial results as "the truth" and putting them into action.

8.3.3
Critical Appraisal

Critical appraisal means assessing to what extent the methodological quality of the evidence supports confidence in the correctness of effect estimates. This quality assessment is commonly performed by systematically checking aspects of a trial design which are particularly susceptible to introduce bias. For example, in trials assessing the efficacy of a treatment the following questions are critical: "Was the randomization process adequate and the sequence of allocation concealed?", "Were the baseline characteristics of intervention and control group comparable?", "Were patients and doctors blinded to intervention and outcome assessment?", "Were relevant outcomes used?" "Were all patients included in the trial accounted for in the results section?", etc. There is a number of instruments and checklists provided by different organizations (e.g. Centre for EbM, Users Guide Interactive [5] or the National Health Service Public Health Resource Unit [6]) that may be used to guide the appraisal process.

The LoE (study design) plus the results of the critical appraisal process may now be used to grade the strength of a recommendation. "Grading" means to point out how strong the underlying evidence in support of a recommendation is. The Oxford Centre for EbM suggests using grades A–D (Table 8.2). This type of grading is often used within evidence based clinical guidelines, signing each guideline statement with a grade of recommendation.

After carefully checking the WOSCOP study in our example, no relevant quality deficiencies could be revealed. The study and its results have to be accepted as valid. Therefore, the result of our critical appraisal and the LoE 1b would lead to a "Grade A" recommendation. Thus, we conclude at this point that there is strong evidence that over the next 5 years the intake of a statin drug will lower the risk of dying of CHD in asymptomatic males with multiple risk factors.

Table 8.2 Grades of recommendation (adapted from the Oxford Centre for EbM).

Grade of recommendation	
A	consistent level 1 studies
B	consistent level 2 or 3 studies *or* extrapolations from level 1 studies
C	level 4 studies *or* extrapolations from level 2 or 3 studies
D	level 5 evidence *or* troublingly inconsistent *or* inconclusive studies of any level

Further details under http://www.cebm.net/levels_of_evidence.

8.3.4
Acting on the Evidence

This step means that useful findings should be implemented into clinical practice. However, when is a "finding" useful? Always then when a high LoE or a high grade of recommendation is present? Certainly not! The final decision to follow the evidence is a clinical decision where not only the statistical significance of an effect should be taken into account, but also its clinical relevance. Therefore, we have to define the relevant effect in our clinical situation. On this basis we are able to assess whether the "finding" (as a result of the critical appraisal) is useful or not. A concept to convert formal evidence to clinical recommendations is the Grading of Recommendations Assessment, Development and Evaluation (GRADE) schema provided by the GRADE working group (http://www.gradeworkinggroup.org).

Let us go back to our example. We can use the effect measures given in the study results section to assess whether the risk reduction after statin intake is clinically meaningful. The WOSCOP study shows a significant 30% reduction of the 5-year risk of dying from CHD in men with hypercholesterinemia – at first view an impressive result. Looking at the absolute figures will reveal that the risk of CHD death was reduced from 2.3% in the placebo group to 1.6% in the statin group (relative risk reduction = 30%). The absolute risk was reduced by 0.7%. This means that if a group of 1000 men with hypercholesterinemia is treated with statins for 5 years, 16 CHD related deaths are to be expected instead of 23 if the group is not receiving the drug. In order to avoid one cardiovascular death 143 patients have to be treated for 5 years [1/0.7% = number needed to treat (NNT)].

However, what advice should be finally given to our patient? Evidence level (1b) and grade of recommendation (A) would favor prescribing a statin. But the possible benefit seems only to be small.

In our situation the doctor and patient together should make the final decision. Statins could be the right choice, but also alternatives such as changing life style (which has similar effect size) are possible. This is a typical clinical situation where a final decision can not be based on external evidence only. All three, combination of external evidence, clinical expertise and circumstance and patients' preferences, have to be taken into respect.

8.3.5
Evaluate your Performance

The last step of the EbM process is one of the most difficult steps. The modified action has to be evaluated in the own clinical context. Are the results really better after following the new evidence? This requires the consequent application of the recommendations (evidence based guideline) with following outcome evaluation.

8.4
Comparing the EbM to Other Approaches of Characterizing Evidence

The approaches of EbM and other approaches to characterize evidence discussed in this volume reveal many parallels. The process of evidence identification (systematic literature search) and critical appraisal (assessment of quality) are quite similar. Adequate study design, strong association, consistent findings, directness, dose–effect relationship and control of confounding will lead to higher-quality assessments in both approaches. Lower quality will be assumed when findings are inconsistent, and wide confidence intervals of effect measures or biases and confounding are present.

Concerning the endpoint definition, EbM differs from the other approaches. EbM is confined to considering clinically relevant endpoints such as morbidity and mortality measures, quality of life, costs, etc. The other approaches focus on (sometimes vaguely defined) adverse health effects including biological effects (often without a clearly established association to health consequences).

A major difference between EbM and the other approaches is the way how evidence is categorized. While EbM uses a sophisticated system of transforming LoE and systematically assessed quality criteria into graded recommendations, most of the other approaches use less formal procedures.

Finally, EbM is a rather conservative and strict approach to assess "effects" in a clinical context. Only if clear evidence and "usefulness" are present, will this evidence be taken into account for patients' further care. This seems to be justifiable for the patients' view as for the healthcare system view. When a new method is to be implemented into the healthcare system this more conservative approach does make sense. There has to be a clear proof that a new intervention is useful and that it does more good than harm (and that the cost are arguable).

For a risk assessment in the field of electromagnetic fields (EMFs) the EbM approach might be too conservative. Whereas EbM would "block" uncertain or small effects, in the field of EMFs small risk should not be missed. Even biological effects with unclear implications for health could hint to clinically relevant harm.

The EbM approach and the other approaches are justifiable in their context. Systematic literature search, identification of bias and confounding, and quality assessment are comparable. An exchange of methods could be helpful for risk assessment in the discussed approaches.

References

1 Sackett, D.L., Rosenberg, W.M., Gray, J.A., Haynes, R.B. and Richardson, W.S. (1996) Evidence based medicine: what it is and what it isn't, *British Medical Journal*, **312**, 71–2.

2 Haynes, R.B., Devereaux, P.J. and Guyatt, G.H. (2002) Physicians' and patients' choices in evidence based practice – evidence does not make decisions, people do, *British Medical Journal*, **324**, 1350.

3 Sackett, D., Haynes, R., Guyatt, G. and Tugwell, P. (1991) *Clinical Epidemiology,*

Lippincott Williams & Wilkins, Philadelphia, PA.
4 Shepherd, J., Cobbe, S.M., Ford, I., Isles, C.G., Lorimer, A.R., MacFarlane, P.W., McKillop, J.H. and Packard, C.J. (1995) Prevention of coronary heart disease with pravastatin in men with hypercholesterolemia. West of Scotland Coronary Prevention Study Group, *New England Journal of Medicine*, **333**, 1301–1307.
5 American Medical Association Unified Service Center (2002) *Users' Guide Interactive. An online tool to guide clinicians in the appraisal and application of evidence into their everyday practice*, American Medical Association Unified Service Center, Chicago, IL [http://www.usersguides.org/] [Retrieved: 20.03.2007].
6 NHS (2006) *Critical Appraisal Tools*, Critical Appraisal Skills Programme (CASP), Public Health Resource Unit, London [http://www.-phru.nhs.uk/casp/critical_ appraisal_ tools.htm] [Retrieved: 20.03.2007].

9
The *IARC Monographs*' Approach to Characterizing Evidence

Vincent James Cogliano, Robert Alexander Baan, Kurt Straif, Yann Grosse, Marie Béatrice Secretan, and Fatiha El Ghissassi

9.1
Introduction

The *IARC Monographs on the Evaluation of Carcinogenic Risks to Humans* is a series of scientific reviews that identify environmental factors that can increase the risk of human cancer. Each *Monograph* includes a critical review of the pertinent scientific literature and an evaluation of the weight of the evidence that the agent can alter the risk of cancer in humans. A distinguishing feature is that each *Monograph* is developed by an international, interdisciplinary Working Group of expert scientists who conducted the original research. Since the programme began in 1971, these Working Groups have involved over 1000 scientists from 51 countries.

The scope of the *IARC Monographs* has expanded beyond an original focus on chemical carcinogens to include complex mixtures, occupational exposures, physical and biological agents, and lifestyle factors. More than 900 agents have been evaluated and approximately 400 have been classified as *carcinogenic, probably carcinogenic* or *possibly carcinogenic to humans*. National and international health agencies use this information as scientific support for their actions to prevent exposure to potential carcinogens.

9.2
Pertinent Data for Carcinogen Identification

Sources of data for identifying potential carcinogens include epidemiological studies, long-term bioassays in experimental animals, and mechanistic and other relevant data. Epidemiological studies can provide uniquely informative results about the response of humans exposed to potential carcinogens and other toxic agents. Among these, cohort and case-control studies are especially useful for determining whether

The Role of Evidence in Risk Characterization: Making Sense of Conflicting Data.
Edited by Peter M. Wiedemann and Holger Schütz
Copyright © 2008 WILEY-VCH Verlag GmbH & Co. KGaA, Weinheim
ISBN: 978-3-527-32048-6

exposure to an agent is causally associated with human cancer. There are, however, inherent limitations to what epidemiology can study. It can be difficult to accurately estimate exposure or to attribute causality to a single factor and epidemiologic studies cannot rule out small risks. In addition, the latent period for cancer implies that many years of human exposure will occur before epidemiological studies can be considered informative.

For these reasons, long-term bioassays in experimental animals are presently the most common means of assessing potential risks to humans. The strengths and limitations of bioassays complement those of epidemiologic studies. Cancer bioassays are based on the scientific assumption that agents causing cancer in animals will have similar effects in humans [1]. The controlled nature of animal experiments means that exposures can be clearly prescribed without most sources of potential bias and confounding. In addition, cancer bioassays in rats and mice can generally be conducted before informative epidemiologic studies are available.

Mechanistic and other relevant data are being used to complement studies in humans and experimental animals. Toxicokinetic studies can allow comparison of absorption, distribution, metabolism and elimination (ADME) across species, although in many cases detailed studies are conducted only in animals. Mechanistic studies attempt to identify the cellular processes leading to tumor development. This knowledge can be used to establish a correspondence between experimental animals and humans and to identify susceptible populations and life stages.

9.3
International Agency for Research on Cancer Evaluations

IARC Monograph evaluations are developed through a series of distinct steps that are described in guidelines known as the Preamble to the *IARC Monographs* [2]. *IARC Monograph* evaluations are developed by a Working Group of expert scientists who conducted the original research. The stepwise process provides insight into the evaluations by revealing the weight given by the Working Group to each line of evidence.

9.3.1
Evaluating Epidemiologic Studies

The epidemiologists on the Working Group consider the studies of cancer in humans and communicate their evaluation in clear terms that can be understood by people outside the field. The epidemiological evidence is characterized using standard descriptors that span a range of levels of evidence [2].

- *Sufficient evidence of carcinogenicity.* A causal relationship has been established between exposure to the agent and human cancer, i.e. a positive relationship has been observed in studies in which chance, bias and confounding could be ruled out with reasonable confidence.

- *Limited evidence of carcinogenicity.* A positive association has been observed between exposure to the agent and cancer for which a causal interpretation is credible, but chance, bias or confounding could not be ruled out with reasonable confidence.
- *Inadequate evidence of carcinogenicity.* The available studies are of insufficient quality, consistency or statistical power to permit a conclusion regarding the presence or absence of a causal association between exposure and cancer, or no data on cancer in humans are available.
- *Evidence suggesting lack of carcinogenicity.* There are several adequate studies covering the full range of levels of exposure that humans are known to encounter, which are mutually consistent in not showing a positive association between exposure to the agent and any studied cancer at any level of exposure. Bias and confounding should be ruled out with reasonable confidence, and the studies should have an adequate length of follow-up. A conclusion of *evidence suggesting lack of carcinogenicity* is inevitably limited to the cancer sites, conditions and levels of exposure, and length of observation covered by the available studies. In addition, the possibility of a very small risk at the levels of exposure studied can never be excluded.

The key questions are whether a causal interpretation is credible and whether chance, bias or confounding could be ruled out with reasonable confidence. To answer these questions, epidemiologists have found useful guidance in a set of factors commonly known as the "Hill criteria" [3, 4].

- *Consistency of the observed association.* Reproducibility of findings across independent studies and independent settings constitutes one of the strongest arguments for causality.
- *Strength of the observed association.* The finding of large risks increases confidence that an association is not due to chance, bias or confounding; a modest risk, however, does not preclude causality and may reflect an agent of low potency, a low exposure level, or an increase in a common disease.
- *Specificity of the observed association.* Generally refers to the association of a single cause with a single effect; current understanding is that many agents cause cancer at multiple sites and many cancers have multiple causes, thus absence of specificity does not preclude causality.
- *Temporal relationship of the observed association.* Causality requires that exposure precede development of the disease; in view of the latent period for cancer, it is important to ascertain whether the studies included adequate follow-up time.
- *Biological gradient (exposure–response relationship).* A clear exposure–response relationship strongly suggests cause and effect; however, the ability to find exposure–response relationships may be diminished by exposure misclassification or a small range of exposure levels in a study.

- *Biological plausibility.* An inference of causality is strengthened by consistency with experimental data that show plausible biological mechanisms; however, a lack of mechanistic data does not preclude causality.
- *Coherence.* Other lines of evidence, e.g. experimental animal studies, toxicokinetic studies, short-term tests and mechanistic studies, may strengthen an inference of causality.
- *Experimental evidence (from human populations).* These include randomized control trials or cases where conditions of human exposure are altered to create a "natural experiment" at different levels of exposure: when a change in exposure leads to a change in disease incidence, this provides strong evidence of a causal association, e.g. the decrease in lung cancer risk following smoking cessation.
- *Analogy.* Evidence for causality can be strengthened by information on an agent's structural analogues.

9.3.2
Evaluating Bioassays in Experimental Animals

In like manner, the toxicologists and pathologists on the Working Group consider the studies of cancer in experimental animals and communicate their evaluation in using the same range of standard descriptors [2]. In this case the key question concerns the breadth and replication of tumor findings, e.g. the induction of tumors in multiple species or in independent studies.

- *Sufficient evidence of carcinogenicity.* A causal relationship has been established between the agent and an increased incidence of malignant neoplasms or of an appropriate combination of benign and malignant neoplasms in (i) two or more species of animals or (ii) two or more independent studies in one species carried out at different times or in different laboratories or under different protocols. An increased incidence of tumors in both sexes of a single species in a well-conducted study, ideally conducted under Good Laboratory Practices, can also provide *sufficient evidence*.
- A single study in one species and sex might be considered to provide *sufficient evidence* when malignant neoplasms occur to an unusual degree with regard to incidence, site, type of tumor or age at onset, or when there are strong findings of tumors at multiple sites.
- *Limited evidence of carcinogenicity.* The data suggest a carcinogenic effect but are limited for making a definitive evaluation because, for example, (i) the evidence of carcinogenicity is restricted to a single experiment, (ii) there are unresolved questions regarding the adequacy of the design, conduct or interpretation of the studies, (iii) the agent increases the incidence only of benign neoplasms or lesions of uncertain neoplastic potential, or (iv) the evidence of carcinogenicity is restricted

to studies that demonstrate only promoting activity in a narrow range of tissues or organs.

- *Inadequate evidence of carcinogenicity.* The studies cannot be interpreted as showing either the presence or absence of a carcinogenic effect because of major qualitative or quantitative limitations, or no data on cancer in experimental animals are available.
- *Evidence suggesting lack of carcinogenicity.* Adequate studies in at least two species are available which show that, within the limits of the test used, the agent is not carcinogenic. A conclusion of *evidence suggesting lack of carcinogenicity* is inevitably limited to the species, tumor sites, age at exposure, and conditions and levels of exposure studied.

9.3.3
Evaluating Mechanistic and Other Relevant Data

Mechanistic and other relevant data cover a wide variety of study designs, including toxicokinetic studies of ADME; mechanistic studies that elucidate genotoxic and other mechanisms of carcinogenesis; data on genetic polymorphisms and other determinants of susceptibility across different populations or life stages; and toxic effects that confirm distribution or biological effects at sites of tumor development. Mechanistic data are not evaluated using standard descriptors, as the rapid evolution of mechanistic study designs does not yet lend itself to fixed classification criteria. Instead, the laboratory scientists on the Working Group who study mechanisms of carcinogenesis characterize the mechanistic data as "weak," "moderate" or "strong" and attempt to determine whether the mechanism that causes cancer in experimental animals is likely to be operative in humans [2].

Consideration of mechanistic data has the potential to improve the analysis of studies in humans and experimental animals. For example, if a sequence of key events leading to tumor development has been established, then observation of these tumor precursors in exposed humans can provide strong support for an agent being a cancer hazard. In experimental animals, mechanistic studies have generally been focused on establishing a correspondence between responses in animals and humans.

9.3.4
Overall Evaluation

These separate evaluations of human, animal, and mechanistic data are combined into an overall evaluation of the agent as:

- *Carcinogenic to humans* (Group 1)
- *Probably carcinogenic to humans* (Group 2A)
- *Possibly carcinogenic to humans* (Group 2B)
- *Not classifiable as to its carcinogenicity to humans* (Group 3)
- *Probably not carcinogenic to humans* (Group 4)

Before mechanistic data were widely available, the overall evaluation was usually determined by the strength of the evidence in humans and experimental animals. That is:

- If there was *sufficient evidence* in humans, the agent would be classified in Group 1 (*carcinogenic to humans*)
- If there was *limited evidence* in humans and *sufficient evidence* in experimental animals, the agent would be classified in Group 2A (*probably carcinogenic to humans*)
- If there was *limited evidence* in humans or *sufficient evidence* in experimental animals, but not both, the agent would be classified in Group 2B (*possibly carcinogenic to humans*)
- If there was *limited* or *inadequate evidence* in experimental animals (and *inadequate evidence* in humans), the agent would be classified in Group 3 (*not classifiable as to its carcinogenicity to humans*)
- If there was *evidence suggesting lack of carcinogenicity* in both humans and experimental animals, the agent would be classified in Group 4 (*probably not carcinogenic to humans*)

Mechanistic data can be pivotal in International Agency for Research on Cancer (IARC) evaluations when the human evidence is not conclusive (i.e. there is either *limited evidence* or *inadequate evidence* in humans). In such cases, mechanistic data can be used to arrive at any of the IARC classifications, including Group 1 (*carcinogenic to humans*). Specifically,

- An agent may be raised to Group 1 with less than *sufficient evidence* in humans if there is *sufficient evidence* in experimental animals and "strong evidence in exposed humans that the agent acts through a relevant mechanism of carcinogenicity". This opens up the possibility of a Group 1 classification for agents that cannot be the subject of epidemiologic studies.
- An agent may be raised from Group 2B to Group 2A with *inadequate evidence* in humans and *sufficient evidence* in experimental animals plus "strong evidence that the carcinogenesis is mediated by a mechanism that also operates in humans".
- Alternatively, an agent may be assigned to Group 2A even without epidemiologic studies or animal bioassays if it "clearly belongs, based on mechanistic considerations, to a class of agents for which one or more members have been classified in Group 1 or Group 2A".
- An agent may be raised from Group 3 to Group 2B with only *limited evidence* in experimental animals if there is also supporting evidence from mechanistic and other relevant data.
- Completing this idea, a Group 2B classification can be equally appropriate when there is *inadequate evidence* in experimental animals (and also *inadequate evidence* in

humans), but there is strong evidence from mechanistic and other relevant data. An agent may be classified in Group 2B solely on the basis of strong evidence from mechanistic and other relevant data.

- In the other direction, strong mechanistic evidence can be used to lower a classification that would otherwise be indicated by the results of cancer bioassays. An agent may be lowered from Group 2B to Group 3 if there is *inadequate evidence* in humans and *sufficient evidence* in experimental animals plus "strong evidence that the mechanism of carcinogenicity in experimental animals does not operate in humans".

- An agent may be lowered from Group 3 to Group 4 if there is *inadequate evidence* in humans and *evidence suggesting lack of carcinogenicity* in experimental animals and this is "consistently and strongly supported by a broad range of mechanistic and other relevant data". This opens up the possibility of a Group 4 classification for agents that have not or cannot be the subject of epidemiologic studies.

Thus the overall evaluation is a matter of scientific judgment, reflecting the weight of the evidence derived from studies in humans, from studies in experimental animals, and from mechanistic and other relevant data. In considering all relevant scientific data, an agent may be assigned to a higher or lower group than the default would indicate [2]. The Working Group strives to achieve a consensus evaluation. The evaluation includes a discussion of the rationale for the conclusions.

9.4 Hazard versus Risk

The above categories refer only to the strength of the evidence that an agent is carcinogenic and not to the extent of its carcinogenic activity or potency [2]. A cancer *hazard* is an agent that is capable of causing cancer under some circumstances, while a cancer *risk* is an estimate of the carcinogenic effects expected from exposure to a cancer hazard.

A distinction between hazard and risk has been made for many years. These concepts were formalized in a risk assessment paradigm that describes risk assessment as a series of distinct steps [5, 6]. *Hazard identification* determines whether exposure to an agent is linked to adverse health effects. *Dose–response assessment* characterizes the relation between the dose of an agent and the incidence of an adverse health effect. *Exposure assessment* determines the extent of human exposure to an agent. *Risk characterization* describes the nature and magnitude of human risk, including attendant uncertainty. Under this paradigm, risk depends on both the existence of a hazard and exposure to that hazard.

The distinction between hazard and risk is important, and the *IARC Monographs* evaluate cancer hazards (despite the historical presence of the word "risks" in the title of each volume of *Monographs*) even when risks are very low at current exposure levels, because new uses or unforeseen exposures could engender risks that are significantly higher. For example, short-term exposure to overhead fluorescent lighting that mimics sunlight may pose a negligible cancer risk, but more intense short-term exposure to the same lighting in sunbeds can pose risks that are more substantial. Another example may be a herb that is generally consumed in such small quantities that it poses little cancer risk, but if that herb is later packaged in concentrated amounts as a dietary supplement, there may be a high risk of cancer. In both these cases, it is important to identify these agents as cancer hazards even though the cancer risk is generally very low under normal circumstances, in order that a cancer hazard would be anticipated in the case of new uses that involve higher exposures, such as sunbeds or dietary supplements.

9.5
Ensuring Impartial Evaluations

As it is important to ensure that evaluations are impartial and not subject to influence by special interests, the IARC strives to identify and avoid real or apparent conflicts of interests in the Working Group members who develop the evaluations. The IARC selects experts on the basis of (i) knowledge and experience, and (ii) absence of real or apparent conflicts of interests. Consideration is also given to demographic diversity and balance of scientific findings and views. A difficulty arises when an expert with relevant knowledge and experience also has a real or apparent conflict of interests. Such experts are invited only when necessary and do not serve as chair, draft text that pertains to the description or interpretation of cancer data, or participate in the evaluations. In this way, a *Monograph* meeting can include the best-qualified experts and the evaluations are developed and written by experts with no conflicting interests [2, 7].

The IARC also strives to keep its Working Groups free from all attempts at interference, before, during and after a meeting. This includes lobbying by interested parties, receipt of written materials from interested parties and offers of meals, drinks, social invitations or other favors offered by interested parties. Attempts at interference outside the meeting are particularly insidious, as they occur outside the view of IARC staff and other participants. Meeting participants are responsible for safeguarding the integrity of their work by resisting all attempts to influence the meeting. To aid them in this responsibility, IARC reminds meeting participants to report all attempts at interference [2, 7].

9.6
Characterizing Evidence in the Future

The future of hazard identification and risk assessment will be one of continuing evolution to reflect changes in the underlying science. Future evaluations will continue to consider mechanistic data to aid in interpreting experimental animal results. When sufficient data are available to identify an agent's mechanisms of carcinogenesis, these data will also be the key to identifying susceptible populations and life stages, including the prenatal and early postnatal periods. Another implication of using mechanistic data will be more carcinogen identifications that are based on scientific inference in the absence of cancer studies in humans or experimental animals.

Acknowledgements

The *IARC Monographs* are supported, in part, by the US National Cancer Institute, the European Commission, the US National Institute of Environmental Health Sciences and the US Environmental Protection Agency. The authors wish to acknowledge the important contributions of administrative staff of the *IARC Monographs* programme: Sandrine Égraz, Helene Lorenzen-Augros, Martine Lézère and Jane Mitchell.

References

1 National Toxicology Program (2002) *Report on Carcinogens*, 10th edn, Department of Health and Human Services, Washington, DC [http://ntp-server.niehs.nih.gov/] [Retrieved: 24.09.2007].

2 IARC (2006) *Preamble to the IARC Monographs. IARC Monographs Programme on the Evaluation of Carcinogenic Risks to Humans*, International Agency for Research on Cancer, Lyon [http://monographs.iarc.fr/ENG/Preamble/CurrentPreamble.pdf] [Retrieved: 24.09.2007].

3 Hill, A.B. (1965). The environment and disease: association or causation?, *Proceedings of the Royal Society of Medicine* 58, 295–300.

4 EPA (2005) *Guidelines for Carcinogen Risk Assessment*, review draft edn, US Environmental Protection Agency, Washington, DC [http://www.epa.gov/ncea/] [Retrieved: 24.09.2007].

5 National Research Council (1983) *Risk Assessment in the Federal Government: Managing the Process*, National Academy Press, Washington, DC.

6 National Research Council (1994) *Science and Judgment in Risk Assessment*, National Academy Press, Washington, DC.

7 Cogliano, V.J., Baan, R.A., Straif, K., Grosse, Y., Secretan, M.B., El Ghissassi, F. and Kleihues, P. (2004) The science and practice of carcinogen identification and evaluation, *Environmental Health Perspectives* 112, 1269–1274.

10
The Swiss Health Risk Approach
Martin Röösli

10.1
Background

In 2000, a new regulation for protection from nonionizing radiation of stationary sources was introduced in Switzerland [1]. The principle of the new regulation was to protect the population from established harmful or annoying effects of nonionizing radiation. In addition, a precautionary approach was applied: given the scientific uncertainty, exposure levels at places where people live for prolonged periods of time should be reduced as much as feasible from a technical and economical point of view. A periodical review of the standard limits was planned.

To obtain the scientific base for a periodic review of the published literature about effects on electromagnetic field (EMF) exposure on human's health, an internet database was established. Relevant publications have been summarized and commented on in German, and have been made freely accessible to the public. The database is called ELMAR and is still maintained (http://www.elmar.unibas.ch). In addition, a systematic review of the scientific literature about health risks from high-frequency EMFs in the radiofrequency and microwave frequency range was carried out [2]. In the framework of this review, the Swiss Health Risk approach was developed. The review was based on studies in humans dealing with exposure in the "low dose range". The term "low dose range" refers to radiation intensities lying below or of the order of the exposure limit values stated in the Ordinance Relating to Protection from Non-Ionizing Radiation (28–61 V/m) [1] or, in the case of mobile telephones, below or at the limit value for the local specific absorption rate recommended by the International Commission for Non-Ionizing Radiation Protection (2 W/kg) [3]. The literature was selected from specialized databases using standardized search criteria, and checked for completeness using review articles and reports. Studies published prior to 31 December 2002 were included in the review. The review has been published as a report in German [2]. A summary of the report was published as a peer-reviewed paper [4]. The target audience for the report was governmental and local authorities dealing with nonionizing radiation as well as scientists and

The Role of Evidence in Risk Characterization: Making Sense of Conflicting Data.
Edited by Peter M. Wiedemann and Holger Schütz
Copyright © 2008 WILEY-VCH Verlag GmbH & Co. KGaA, Weinheim
ISBN: 978-3-527-32048-6

interested lay persons. It was intended to update the report regularly. The first update was published in 2004 [5] and additional updates are scheduled.

10.2
Aims

The Swiss Health Risk Approach was developed to systematically evaluate the scientific evidence of health effects from high-frequency radiation. The approach was developed to apply on experimental studies in humans and on epidemiological studies. It focused on three aspects to evaluate health risks:

1. *Evidence:* to evaluate the strength of evidence on a gradual scale for a given effect
2. *Relevance to health:* to assess of the impact on health
3. *Exposure:* to assess the respective exposure threshold

10.3
Approach

10.3.1
Evidence Rating

For the rating of the evidence, we adapted the established evidence scale of the International Agency for Research on Cancer (IARC) (see Chapter 9). This five-step scale has been used since 1972 to evaluate the carcinogenicity of more than 800 environmental agents and exposures. In order to make the scale applicable also to noncancer endpoints in relation to EMF exposure, the definitions were modified as follows:

- *Established.* An effect is regarded as established if it meets stringent scientific criteria, i.e. is replicated several times in independent investigations, is not in contradiction with other research results and a plausible biological model exists.

- *Probable.* An effect is classified as probable if it has been found repeatedly and with relative consistency in independent studies. The studies concerned must be of a sufficiently high quality to exclude other factors with a large degree of certainty. No plausible causation mechanism is known.

- *Possible.* Effects are regarded as possible where they occur sporadically in the studies. However, the results are not entirely consistent and could be the result of methodological weaknesses. The scientific evidence is corroborated by case reports.

- *Improbable.* There are no indications of an association, but multiple indications of its absence. No theoretically plausible biological model exists.

- *Not assessable*. The data is too scant for an assessment to be made. While isolated evidence exists, this is often contradictory. The methodology of the studies concerned is regarded as insufficient to permit conclusions to be drawn.

10.3.2
Relevance to Health

The relevance to health of all effects was classified into three groups:

- *Serious*. The effect leads to a drastic reduction in the quality of life. It constitutes a threat to life and reduces life expectancy. This category includes all cancerous diseases, genotoxic effects, teratogenicity, stillbirths and increased mortality.
- *Reduced well-being*. While the effect does not represent a direct threat to life, it significantly curtails the quality of life and/or well-being. This category includes unspecific health symptoms such as headaches, insomnia, psychic symptoms, electromagnetic hypersensitivity and microwave hearing.
- *Relevance to health unknown*. The effects are physiologically measurable but lie within the normal variability range of healthy individuals. Such effects do not represent a risk to health *per se* and, since they are normally not perceived, do not lead to a reduction in quality of life. It is not known whether they represent a risk to health in the long term. This group includes fluctuations in the hormone, immune and cardiovascular systems, variability in the electroencephalogram and changes in the perception and processing of stimuli.

10.3.3
Exposure Levels

For effects classified as established, probable or possible, an exposure threshold for their appearance was estimated based on the results of the available studies. To enable the results for mobile telephones and stationary transmission installations to be compared, the different dose metrics had to be standardized. Specific absorption rate per 10 g (SAR_{10}) was chosen as the respective dose metric. While some reports specify SAR_{10} directly, for others it had to be estimated from the incident EMF.

Two exposure groups were defined: 20 mW/kg to 2 W/kg and "in the range of the Swiss installation limit value". The later refers to a SAR below 20 mW/kg. In addition, it was declared which exposure sources were investigated in the available studies.

10.3.4
Summary Scheme

Table 10.1 gives an overview about the health risk assessment carried out in 2002. At that time, brain activity, shortened reaction times, nonspecific symptoms and altered sleep phases related to mobile phone exposure were judged to be probable. Possible effects were considered to be brain tumors and electromagnetic hypersensitivity

Table 10.1 Summary of the evidence for high-frequency radiation effects on health at low dose levels by end of 2002.

Evidence	Effect			Exposure source	Exposure threshold
	Serious	Detriment to well-being	Relevance to health unknown		
Established (consistent findings)		interference effects on implanted medical devices		electronic appliances (e.g. mobile telephones)	
		microwave hearing		radar installations	energy flux density per pulse >20 mJ/m²
Probable (multiple indications)			brain activity	mobile telephones	20 mW/kg to 2 W/kg
			shortened reaction times	mobile telephones	20 mW/kg to 2 W/kg
		unspecific symptoms (headaches, fatigue, problems of concentration, disquiet, sore skin, etc.)		mobile telephones	20 mW/kg to 2 W/kg
Possible (isolated indications)		sleep quality	sleep phases	mobile telephones radio transmitter	20 mW/kg to 2 W/kg in the region of the installation limit value
		electromagnetic hypersensitivity		mobile telephones	20 mW/kg to 2 W/kg

10.3 Approach

Assessment	Effect	Source	Threshold
	leukemia lymphomas	TV and radio transmitters	in the region of the installation limit value
	brain tumors	mobile telephones	20 mW/kg to 2 W/kg
	mortality	mobile telephones	
Improbable (multiple indications of absence of the effect)	other types of tumor		
Not assessable (insufficient data)	hormone system	various	
		various	
	immune system	various	
	high blood pressure, variable pulse, electrocardiogram	radio transmitters	
	mental symptoms	various	
	unspecific symptoms (insomnia, headaches, etc.)	mobile telephone base stations	
	stillbirth	diathermal appliances	
		exposure at the workplace	
	genotoxicity	various	
	breast cancer	mobile telephones	
	eye tumors		
	tumors of the testicles	radar guns	

The figures for the exposure threshold (given in mW/kg and W/kg) are only intended as a rough guide. They refer to the maximum local SAR_{10} occurring in the body [3].

related to mobile phone use as well as leukemia and impaired subjective sleep quality associated with exposure to emissions from TV and radio transmitters.

10.4
Discussion

10.4.1
Gradual Rating of the Evidence

A strength of the Swiss Health Risk Approach is the gradual rating of the evidence. In the field of EMF research with little established effects, a gradual scale is more informative than one that only distinguishes between established and not established effects. One main challenge was the labeling of the five evidence categories. We adapted the wording of the IARC classification. A survey has shown a wide range of interpretation for the labels of the five evidence categories [6]. In particular, the label "possible" meant for some individuals that it is very unlikely, but cannot be completely ruled out, for others it meant that an effect is likely to occur (see also Chapter 14).

There was also another type of misunderstanding of the labels. The labels were intended to rate the level of evidence concerning the whole research for one certain effect. For instance, it was judged to be "possible" that sleep quality in proximity to TV or radio transmitters may be altered, because such effects "occur sporadically in studies. However, the results are not entirely consistent and could be the result of methodological weaknesses. The scientific evidence is corroborated by case reports". From a scientific point of view, this is a relatively weak statement that the issue is not resolved. Some lay people did not understood the label as a rating of the total research evidence, but as a rating of their individual risk to develop sleep disorders if living close to a transmitter. However, there is a fundamental difference between rating research evidence and rating the individual risk. For instance, there is established evidence for an association between cigarette smoking and lung cancer. Still, not every smoker develops lung cancer and thus a smoker does not necessarily develop a lung cancer, but it is "possible". Reversely, if a person believes that it is possible to develop sleep disorders in the proximity of a transmitter, this means inherently that there is established evidence for an association between exposure to a transmitter and sleep disorders.

10.4.2
Source-specific Evaluation

Choosing the relevant exposure measure is difficult in the absence of a known biological mechanism in the low dose range. Sure, the intensity of the exposure is the primary candidate. However, frequency, type of modulation, duration and timing of exposure or other characteristics might also be of importance and may modify potential effects.

It is also challenging to compare exposure levels from sources close to body (such as from mobile phones) with far field sources (fixed site transmitter). Sources close to the body result in a highly localized exposure (e.g. of the head), whereas fixed-site transmitters generate a more homogenously distributed exposure over the whole body. Thus, the question arises as to compare maximum absorption rate (SAR), organ-specific SAR or whole-body SAR. For the latter, exposure differences between near- and far-field sources would be smaller compared to the maximum SAR.

There are also different ways to allow for differences in exposure duration and timing in a risk assessment. For example, exposure from fixed-site transmitters at the place of residence is rather continuous compared to a typical intermittent mobile phone use pattern. Evaluating cumulative or mean levels produces different results than evaluating maximum levels.

To overcome these difficulties, we decided to summaries the evidence for each of the sources separately. In doing so, we did not need to make assumptions about the underlying biological mechanism. If there were systematic different effects for various sources, one could systematically evaluate the causes for it.

10.4.3
Lack of Data

It is typical for the EMF research field that a very wide range of health outcomes has been hypothesized to be related to EMF exposure. Some of the hypotheses were repeatedly investigated (e.g. brain tumor risk and mobile phone use). Other effects have only been rarely investigated, e.g. unspecific symptoms and long-term exposure in the low dose range (e.g. from a fixed-site transmitter). On the one hand, lack of studies can be an indicator that scientists do not find an association to be plausible at all and thus not worth the time and money to investigate it. On the other hand, lack of data can be the result of methodological challenges or of a newly introduced technique. In this case one might consider that "absence of evidence is not evidence of absence of risk". For that reason, a category "not assessable" is useful and informative, since it points at questions researchers have not tried to tackle yet.

10.4.4
Publication Bias

Closely related to the lack-of-data issue is the issue of publication bias. If a scientist observes an unexpected association between an outcome and EMF exposure, the likelihood of publication is relatively high even if there is no known underlying mechanism for it. On the other hand, it is very hard to publish an absence of association without an established mechanism or hypothesis why there should be an association. As a consequence, for some health outcomes there are a few methodological limited studies (sometimes called pilot studies, preliminary analyses, etc.) pointing to an association, but no studies proving the absence of this specific association. Such effects would fit our evidence definition of "possible". Maybe, in the long run, negative studies will also be published if they can refer to previous unexpected observations and contradict them. If so, the effects would be downgraded to "improbable". It is conceivable that the evidence for some reported effects of nonionizing radiation starts with "not assessable" followed by "possible" and ending finally in the category "improbable" when the research moves forward.

There is no simple solution to the problem of evidence rating in the absence of methodological sound studies.

10.4.5
Rating of the Study Quality

Information on study quality is important for the conduct of unbiased systematic review. However, the assessment of the methodological quality of a study is closely intertwined with the quality of reporting [7]. Incomplete and inadequate reporting of research hampers the critical appraisal and appropriate interpretation of research findings. Deficiencies in the reporting of randomized controlled trials led to publication of the CONSORT statement [8]. A substantial part of the literature on the risk of EMFs is of observational nature. Observational research serves a wide variety of purposes using many different study designs. As a result, the development of reporting guidelines is more challenging and has not been established so far. Currently, a collaborative initiative of epidemiologists, methodologists, statisticians, researchers and editors is developing the STROBE Statement (STrengthening the Reporting of OBservational studies in Epidemiology: http://www.strobe-statement. org), which may result in a more consistent reporting of observational studies in the future.

In our approach, strengths and limitations of each study were qualitatively discussed. The overall rating of the evidence took account of the study quality in a nonformal way. Deficient studies that were widely recognized by the public but did not fulfill basic scientific criteria were discussed, but were given no weight in the evidence rating.

10.4.6
Meta-analyses

Meta-analyses allow a systematic evaluation of exposure–response associations and thus, are more informative compared to narrative descriptions. In the field of EMF research, meta-analyses are often not feasible because only a few studies per outcome are available. Even if several studies investigated the same outcome, there are still methodological differences resulting in differing data types that cannot be directly pooled. A typical example is studies on brain activity: Although several studies are available, reported information cannot be directly pooled.

One has also to take into account that meta-analyses address only the statistical uncertainties. In the case of observational research potential biases can be of larger concern compared to statistical uncertainty. For instance, in 2002 there were several studies available investigating hematopoietic and lymphatic malignancies in relation to distance from TV and radio transmitters. Pooling of the exposure–response association had been possible for this specific outcome and exposure situation, and would have resulted in a relatively precise effect estimate. However, this pooled effect estimate would not have reflected potential biases in these studies. There were indications of a cluster problem, meaning that studies were conducted primarily in areas where an *a priori* eye-catching increase of leukemia was observed. Additional bias in these studies may also arise from lack of controlling for confounding factors (e.g. socioeconomic standards) or uncertainties in exposure assessment. For that reason we abandoned from a formal meta-analyses in the Swiss Health Risk Approach.

10.5
Conclusions

The EMF health risk research field is characterized by a wide variety of outcomes and exposure sources investigated by means of different study designs. The Swiss Health Risk Approach was developed to systematically evaluate studies in humans exposed to EMF in the low dose range. The approach rates three different aspects: level of evidence, health relevance and the exposure threshold. Our approach was developed to evaluate empirical evidence and considered the absence of established biological mechanisms in the low dose range by rating the evidence for each of the exposure sources separately. The quality of studies was rated in a non formal way. A future risk assessment may be based on a more formal approach. However, such an approach has yet to be developed and, given the heterogeneous research in the field of EMFs, the applicability has to be proven.

Acknowledgments

Many thanks to Charlotte Braun-Fahrländer and Regula Rapp who contributed to the development of the Swiss Health Risk Approach. Thanks also to Kerstin Hug for valuable and inspiring discussions about the approach. I thank Jürg Baumann for helpful comments on this text.

References

1 Swiss Government (2000) *Ordinance Relating to Protection from Non-ionizing Radiation (ONIR)* [http://www.bafu.admin.ch/elektrosmog/01079/index.html?lang=en&download=NHzLpZig7t,lnp6I0NTU042l2Z6ln1ad1IZn4Z2qZpnO2Yuq2Z6gpJCDe4B,fmym162dpYbUzd,Gpd6emK2Oz9aGodetmqaN19XI2IdvoaCVZ,s-.pdf] [Retrieved: 16.04.2007].

2 Röösli, M. and Rapp, R. (2003) *Hochfrequente Strahlung und Gesundheit*, Bundesamt für Umwelt, Wald und Landschaft (ed.), Umwelt-Materialien Nr. 162, Bern [http://www.bafu.admin.ch/php/modules/shop/files/pdf/php1Rr5fL.pdf] [Retrieved: 16.04.2007].

3 ICNIRP (1998) Guidelines for limiting exposure to time-varying electric, magnetic,

and electromagnetic fields (up to 300 GHz), *Health Physics*, **74**, 494–522.

4 Röösli, M., Rapp, R. and Braun-Fahrländer, C. (2003) Radio and microwave frequency radiation and health – an analysis of the literature [Hochfrequente Strahlung und Gesundheit – eine Literaturanalyse], *Gesundheitswesen*, **65**, 378–392.

5 Hug, K. and Rapp, R. (2004) *Hochfrequente Strahlung und Gesundheit*, Bundesamt für Umwelt, Wald und Landschaft (ed.), Umwelt-Materialien Nr. 162 Nachtrag A, Bern [http://www.bafu.admin.ch/php/modules/shop/files/pdf/phpAqrbDY.pdf] [Retrieved: 18.04.2007].

6 Thalmann, A.T. (2005) *Risiko Elektrosmog. Wie ist Wissen in der Grauzone zu kommunizieren?*, Beltz-Verlag, Weinheim.

7 Huwiler-Muntener, K., Juni, P., Junker, C. and Egger, M. (2002) Quality of reporting of randomized trials as a measure of methodologic quality, *Journal of the American Medical Association*, **287**, 2801–2804.

8 Moher, D., Schulz, K.F. and Altman, D.G. (2001) The CONSORT statement: revised recommendations for improving the quality of reports of parallel-group randomized trials, *Annals of Internal Medicine*, **134**, 657–662.

11
Procedures for Characterizing Evidence: German Commission on Radiation Protection (Strahlenschutzkommission)
Norbert Leitgeb

11.1
Introduction

Scientific risk assessment is challenged by the widespread public myth that safety could be guaranteed and lack of any risk proven if only we tried hard enough. However, this myth ignores a basic scientific dilemma: for fundamental scientific reasons and as a general principle it is impossible to scientifically prove the nonexistence of an effect. Whatever effort might be undertaken, whatever amount of investigation made, at any time it could be argued that effects could have been seen if we tried even harder and longer, and if not at present, their evidence could be expected in the future. Therefore, proof of lack of any risk cannot be achieved, and both scientists and the public have to deal with a certain degree of uncertainty.

The discussion on potential adverse effects of environmental electromagnetic fields (EMFs) is still characterized by different positions of public risk perception and conclusions of bodies responsible for radiation protection. Apart from individual parameters influencing risk perception *per se*, this is caused by differences in:

- Understanding terminology
- Weighting the potential health-relevance of reported effects
- Weighting uncertainty of existing knowledge
- Applying precautionary measures

For health risk assessment, it is necessary to weigh evidence of studies and the relevance of reported effects to the health of humans because

- Not every effect can be considered to be relevant to health
- No single scientific investigation, even of outstanding quality, can be considered to be powerful enough to provide sufficient evidence
- Not every study meets scientific quality criteria

- Not every conclusion of authors is justified by the data
- Evidence needs additional support by already existing knowledge or independent replication and confirmation
- Results gained using nonhuman subjects cannot be extrapolated to humans without adequate models
- Exposure conditions of a study cannot be extrapolated to other situations without adequate models

To account for this, the German Commission on Radiation Protection [Strahlenschutzkommission (SSK)] introduced different degrees of interaction using the hierarchical terms *physical interaction, biologic effect, biological reaction* and *health-relevant effect* using the following definitions [1]:

- *Physical interactions* do not necessarily lead to measurable effects. They could also be just postulated based on established knowledge (such as forces acting on charged particles in the presence of an electric field). Temperature increments or changes of the membrane potentials can be lower than the accuracy of the available measurement equipment. It is typical for such interactions that they do not cause active biologic responses.
- *Biologic effects* are biologic changes as a consequence of physical interactions and might be passive, such as transient local changes of the membrane potential, or active, such as the release of a single action potential by a nerve cell or activation of local thermoregulation by local widening of a blood vessel.
- *Physiologic reactions* can be caused by biologic effects that are strong enough to activate systemic regulatory responses such as thermoregulation. They are not necessarily health-relevant, but might impair well-being or performance.
- *Health-relevant effects* are physiologic reactions exceeding normal physiologic variation range such as core temperature increases above $1\,°C$. They can cause adverse reactions such as annoyance, relevant performance degradation or impair bodily integrity.

11.2
Assessment of Scientific Evidence

The assessment of the existing body of literature is summarized in Tables 11.1 and 11.2. It can be seen that in the extremely low-frequency (ELF) range evidence for an EMF health risk in regard to leukemia of children and neurodegenerative diseases of adults has been classified as justified scientific suspicion. In the radio-frequency (RF) range evidence for biological endpoints including cancer is not considered strong enough to trigger justified scientific suspicion. The literature of most biological endpoints leads to (just) scientific indications. These are considered to indicate a potential causal link of EMF to several biological endpoints. Overall, the SSK considers more scientific research necessary to clarify these open questions. Consequently, it supported the decision to start with an extensive research program on mobile telecommunication fields.

Table 11.1 Assessment of scientific evidence for health-relevant links of ELF EMFs with biological endpoints.

Biological endpoint	Established evidence	Justified scientific suspicion	Scientific indication
Childhood leukemia		×	
Neurodegenerative disease		×	
Cardiovascular system			×
Melatonin (animal)			×
Central nervous system and cognitive function			×
Sleep			×
Psychic effects			×
Electromagnetic hypersensitivity			×

Table 11.2 Assessment of scientific evidence for health-relevant links of RF- EMFs with biological endpoints.

Biological endpoint	Exposure condition	Established evidence	Justified scientific suspicion	Scientific indication
Lymphoma	PB			×
Cancer (animals)	PB			×
Molecules and membranes	PB			×
Calcium concentration	PB			×
Behavior (animals)	PB			×
Electroencephalogram, sleep	PB			×
Cognitive function	PB			×
Blood–brain barrier	PB			×
Blood parameters, immune system	WB			×

PB, partial body exposure; WB, whole-body exposure.

11.3
Relevance to Human Health

The health relevance of scientific studies needs to be evaluated with regard to their relevance to humans. Studies on biological interactions are frequently made with different objects *in vitro* and/or *in vivo*, ranging from molecules, single cells, isolated tissues, living invertebrates, vertebrates and mammals. It is essential to check whether at all or under which restrictions reported effects can be extrapolated to equivalent human exposure situations and human physiological reactions. Such extrapolations need to be based on two different aspects:

- Physical aspects that are governed by physical laws of interaction, such as the induction law for alternating magnetic fields. These are necessary to account for physical differences of investigated objects in relation to humans such as size and structure

- Biologic aspects to account for different physiology, such as circadian activity, thermoregulation or exposure duration relative to lifespan.

Apart from this, physical and biological knowledge is required to allow extrapolation from one exposure situation to another, such as from one frequency to another or to other signal signatures or pulsations. Another challenge is extrapolating exposure durations to humans in regard to the different lifespans of investigated species. It is a yet unsolved problem how to assess biologic responses in the case of chronic exposures to weak fields. While established knowledge does not indicate any accumulative effect that could lead to long-term summation of potential small changes, the hypothetical possibility remains.

The SSK does not consider hypothetical risks to be an adequate basis for mandatory precautionary actions. Examples are epidemiological studies indicating a statistical association between childhood leukemia and power lines. This finding could not be supported by other scientific approaches such as laboratory studies nor explained by known interaction mechanisms. In agreement with other national and international radiation protection bodies such as the UK Health Protection Agency, the International Commission on Non-Ionizing Radiation Protection (ICNIRP) or the World Health Organization, the SSK concludes that epidemiological studies alone without support by laboratory studies or explanatory interaction mechanisms are not sufficient to consider a leukemia risk to children due to ELF magnetic field exposure. However, taking into account the mutually supporting epidemiological results and the classification of magnetic fields as a potential carcinogen class 2B made by the International Agency for Research on Cancer [2], the SSK classifies these findings as scientific indication of a potential health risk below existing limits that merits further investigations, but as insufficient to change limits.

11.4
Weight of Evidence

Assessing scientific quality and rigidity of studies is an overarching issue. With regard to study quality, the SSK sticks to the accepted quality criteria for scientific research such as objectivity, methodology, causality and reproducibility. An important quality indicator is considered the publication in a peer-reviewed scientific journal. Anecdotal reports on patients frequently do not meet minimum requirements of quality, objectivity and reproducibility. Therefore, they are not considered adequate to conclude on causality. However, if reported following the basic guidelines accepted in medicine for case reports they could trigger scientific research such as investigating self-reported electromagnetic hypersensitivity and its potential link to EMFs.

SSK classifies evidence into four categories: *scientific evidence, justified scientific suspicion, scientific indication* and *inadequate data*.

- *Scientific evidence* is assumed if a relationship to the exposure is shown and reproduced by scientific studies performed by independent groups, and causality is supported by the existing overall scientific knowledge.

- *Justified scientific suspicion* is assumed if results of confirmed scientific investigations show a relationship to the exposure, but the overall scientific knowledge does not sufficiently support causality. The degree of suspicion depends on the number and consistency of the existing scientific studies.

- *A scientific indication* is assumed if single studies indicate a relationship to the exposure, but lack support by independent investigations and the overall scientific knowledge.

- *Inadequate data* are given in case of existing single studies of contradictory outcome and/or performed with questionable quality or limited power to conclude on causality.

The SSK is aware that assessment of scientific evidence might be biased by subjective judgments. Therefore, discussions with other stakeholders are organized. Apart from this, judgments of other expert groups involved in scientific health risk assessment are recognized.

It is taken into account that reported scientific results may be uncertain for various reasons. Living beings are not machines and never react in the same way. Therefore, investigations need to be made on a sufficient number of individuals. However, this may have not been achievable, e.g. simply due to restricted financial funds. In case of weak interactions the effects may be small. Therefore, if detectable at all, cofactors may play a dominant role, mask an effect, if existent, and lead to false conclusions. These are the reasons why single studies are not considered powerful enough to provide established evidence and why independent replications are needed.

Apart from this, not every study may have been performed properly. Therefore, it has to be checked whether the experimental setup had been adequate:

- The exposure conditions were sufficiently controlled and quantified
- The preparation of the biological object and the used assay were suited for the investigated endpoint
- The number of investigated samples was sufficient
- The statistical test methods were adequate
- The evaluation was performed thoroughly and accurately
- The drawn conclusions justified

To decide upon this, it is necessary that a study has been documented properly and in sufficient detail. Therefore, documentation is considered an important quality and acceptance criterion.

11.5
Multidisciplinary Assessment

The number and diversity of studies are far too large to allow assessment by one single expert. Therefore, groups and committees are required composed of experts with different expertise, e.g. in medicine, biology, biophysics, engineering, statistics and epidemiology, to assess the scientific quality and relevance of the study outcome. The task is complicated by the complexity of investigated biological objects and endpoints. Assessment is like putting pieces of a scientific puzzle together to get an overall view on health relevance and potential risks rather to pinpoint a single study to base conclusions on it. Apart from the methodical approach to characterize evidence, it is important to define which level of evidence is needed for which kind of action. This is intrinsically tied to the objective of protection and the level of risk.

11.6
Regulations

For mandatory requirements it is the responsibility of politicians rather than scientists to define the necessary level of evidence. Due to the rules of a constitutional state the evidence for setting legal limits need to be sufficiently established to convincingly justify legal requirements. Therefore, legally defined mandatory limits need to be based on established scientific evidence and are aimed at preventing adverse health risks rather than from any reported biologic effect.

In the German ordinance to the federal emission protection law (26. BImSchV), limits of exposures to EMFs are set in terms of field quantities aiming at protecting the general population from EMFs. This ordinance refers to emissions from stationary equipment for RF transmission (10 MHz to 300 GHz) and/or 50- and 16.66-Hz electric power transformation and transport. Field quantities equal ICNIRP reference levels [3]. However, deviating from ICNIRP's recommendations, 2-fold higher values are permitted for short-term exposures (less than 1.2 h) or for small areas outside buildings. EMF limits for occupational exposures are defined in the Employers Association's standard BGV B11 following ICNIRP's concept of basic limits and reference levels.

11.7
Precautions

While the required level of evidence is high for legal measures, the SSK commits itself to the precautionary principle as laid down in the report of the European Commission [4] justifying actions even in case of uncertainty provided they are proportional to the chosen level of protection, nondiscriminatory in their application, consistent with similar measures already taken and reasonable with regard to the cost/benefit ratio of doing or leaving actions.

Therefore, uncertainty needs to be put into perspective. On the one hand, risks from ELF magnetic fields and RF EMFs gain high public awareness. On the other hand, health risks below existing limits still remain hypothetical and so does the estimated number of potential affected cases. Considering the fact that in Germany only about 0.7% of children's sleeping rooms are exposed to 50-Hz magnetic fields above 0.3 µT [5], based on the association derived from epidemiological children leukemia studies a worst case estimate would result in about four leukemia cases per year potentially attributable to elevated magnetic field levels if risk factors from epidemiological meta studies would be applied. Since only 0.3% of such exposures are caused by power lines, the effectiveness of field reduction measures would hardly justify emission limits in particular if restricted to these sources only and ignoring the dominant role of other sources.

Putting risks into perspective would require concentrating first on other risk factors like high exposures to ultraviolet (UV) radiation from the sun or solaria which would merit much more attention since they are already shown to be causally linked with almost 100 000 skin cancer cases per year with thousands of them ending lethally. Therefore, with regard to radiation protection there is an important difference between electromagnetic fields and UV radiation. It remains a challenge to convince the general population to recognize the different severity of risks and to do the most efficient things first, which in this example is changing the habit in exposing children and oneself to UV.

The SSK assigns exposure limits a lighthouse function aiming at timely indicating what levels should be avoided rather than encouraging the industry approaching them as close as possible. This is a consequence of both a precautionary approach and the necessary longsighted view assuring compliance with limits also in the future with additional electric appliances and field sources to come. Therefore, the SSK recommends that overall exposure should stay well below limits. It is recommended to apply the general principle of common sense, and reduce and minimize EMF exposures as far as reasonably achievable.

Following its plea to avoid unnecessary EMF exposure, the SSK requests electric appliances manufacturers to increase their awareness that reduction of EMF emissions could be achieved even with existing technologies without compromising device performance at no or only little additional costs, if minimizing strategies are already implemented in the planning and design phase. Therefore, emission minimization is recommended to be made a criterion for the design and installation of electric appliances, and information on EMF emissions including markings requested to allow informed purchase.

Since no single source should make full use of the tolerated exposure range it is not the question whether at all, but to which extent a single source should be allowed to contribute to environmental exposures. Consequently, the SSK opposes the existing European device standards allowing single electric appliances to emit EMFs up to reference levels and this already under favorable conditions such as free running operation distant to the body in spite of worst-case intended use such as nominal load conditions and use close to the body. In contrast to that, in 2007 the SSK issued a recommendation asking device manufacturers to carefully consider additional sources and environmental fields if EMF emissions from their device exceed one-third of the reference level, and to include in the instructions for use warnings and instructions how to meet exposure limits if emissions exceed two-thirds of the reference level.

It is evident that large-area exposing field sources such as power lines should not be allowed to emit field levels up to the exposure limit in residential areas such as houses or apartments where other field sources are commonly used in daily life. There is an ongoing debate whether source-specific limits should be specified and, if so, on which rational they should be based. The options are, on the one hand, to apply pure technical estimations on potential superposition with simultaneous EMF-emitting appliances and, on the other hand, to add health-related precaution-based considerations such as limiting emissions below values of about 300 nT that are associated with increased child leukemia risk.

11.8
Electromagnetic Interference

The SSK is aware of the fact that the existing EMF exposure limits are not sufficient to protect from adverse electromagnetic interference (EMI) with implanted electronic medical devices such as pacemakers or defibrillators. Facing the ongoing increase of implants in Germany (e.g. from 21 000 units in 2001 to 65 000 in 2005) and even lethal EMI consequences (up to 36% of sudden-death pacemaker patients exhibited a premortem tachyarrythmia), adequate preventive measures are justified. This is a prime challenge to implants manufacturers and to patients to avoid situations of EMI risk. However, the SSK recommends keeping EMF emissions below pacemaker interference levels in particular if sources are invisible such as behind walls or buried (e.g. power cables).

11.9
Conclusions

After assessing the body of scientific literature, the SSK concludes that there is not sufficient evidence to justify changes of the radiation protection concept proposed by ICNIRP and the EU council or to justify implementation of additional precautionary limits. However, for several reasons single sources should not emit EMFs up to exposure limits. If source-specific limits are specified they would not compromise existing exposure limits, but reasonably supplement them.

As a general rule, the SSK recommends reducing unnecessary EMF exposure in particular if this is possible at limited or no additional costs.

References

1 SSK (2001) *Grenzwerte und Vorsorgemaßnahmen zum Schutz der Bevölkerung vor elektromagnetischen Feldern – Empfehlungen der Strahlenschutzkommission (Berichte der SSK, Heft 29)*, Urban & Fischer Verlag, Jena.

2 IARC (2002) *IARC Monographs on the Evaluation of Carcinogenic Risks to Humans: Volume 80. Non-ionizing Radiation, Part 1: Static and Extremely Low-frequency (ELF) Electric and Magnetic Fields*, International Agency for Research on Cancer, Lyon.

3 ICNIRP (1998) Guidelines for limiting exposure to time-varying electric, magnetic, and electromagnetic fields (up to 300 GHz), *Health Physics*, **74**, 494–522.

4 CEC (2000) *Communication on the Precautionary Principle, COM (2000) 1*, Commission of the European Communities Brussels [http://ec. europa.eu/environment/docum/20001_en.htm] [Retrieved: 27.09.2007].

5 Schüz, J., Grigat, J.P., Stormer, B., Rippin, G., Brinkmann, K. and Michaelis, J. (2000) Extremely low frequency magnetic fields in residences in Germany. Distribution of measurements, comparison of two methods for assessing exposure, and predictors for the occurrence of magnetic fields above background level, *Radiation and Environmental Biophysics*, **39**, 233–40.

12
Lessons from the California Electromagnetic Field Risk Assessment of 2002
Raymond Richard Neutra

12.1
Introduction

During the 1990s I had the responsibility of overseeing an evaluation of the potential health effects from power frequency magnetic fields for the California Department of Health Services [1]. Since we needed to format our conclusions so as to be used in a quantitative decision analysis that required a degree of certainty that magnetic fields could cause some degree of excess risk, we needed to go beyond the International Agency for Research on Cancer (IARC) classification [2] and we needed to provide a transparent rationale for the degree of certainty that we claimed. Therefore, with the help of Professor Robert Goble and colleagues of Clark University, we developed Risk Evaluation Guidelines that were submitted for public comment, revised and then approved unanimously by an outside Science Advisory Panel [3]. We then carried out the risk evaluation adhering to the pre-agreed upon guidelines. This was released in draft form and revised before final release in June 2002 [4]. Sunshine laws and the influence of a stakeholder's advisory committee led to a very transparent process, and to criteria for the selection of our science advisors that precluded financial conflict of interest or biases from scientific prejudgment.

Since that time I have done considerable reading about the structure of argument, and have reviewed the components that Bradford Hill [5] and Koch [6] had recommended as the basis for causal arguments in medicine. If I had known in 1998 what I know now, the Electromagnetic Field (EMF) Risk Evaluation Guidelines would be better justified even though they would not be substantially different.

12.2
Policy Questions and Questions about Causal Claims and Arguments

Individual citizens, legislators and regulators concerned with health hazards ought to be asking the following question:

The Role of Evidence in Risk Characterization: Making Sense of Conflicting Data.
Edited by Peter M. Wiedemann and Holger Schütz
Copyright © 2008 WILEY-VCH Verlag GmbH & Co. KGaA, Weinheim
ISBN: 978-3-527-32048-6

How certain must we be of how many cases of disease from this putative hazard before we start exploring the pros and cons of alternative courses of action, alternative mitigations or adopting cheap or expensive mitigation measures?

They should not be asking scientists to answer that question, which after all depends on the balance of stakeholder interests, costs and ethical worldviews, and not just on scientific certainties. What these decision makers, and indeed judges in legal proceedings, *should* ask the scientists are these questions:

- To what degree are you willing to certify that this agent can cause some increased incidence of disease?
- How many cases of disease do you claim could be caused by the typical range of exposures to this agent?
- What is the quantity, quality and the direction of the evidence upon which you ground your claim?
- What general rules do you use for deciding what is and what is not acceptable evidence?
- What are the general inferential rules ("warrants") that you use to serve as a basis for moving from an observed pattern of evidence to justify your claims of scientific certainty?

12.3
Bradford Hill's and Koch's Questions

Bradford Hill's 1965 article [5] is relevant to this *last* question. Scientific certainty of a hazard (or "willingness to certify") can range from "virtually certain that agent X is not a hazard" to "virtually certain that agent X is a hazard". As Toulmin [7] pointed out, when scientists talk about the "probability" that an agent can cause disease or that carbon dioxide emissions will lead to global warming, they are neither talking about some objective property of the external world (as they might by making assertions on the prevalence of "obesity") nor are they making assertions about their state of mind. They are making assertions that they have gone through some kind of orderly scientific argumentation to make a claim about their willingness to certify that the agent is a hazard. "Willingness to certify" is perhaps the most transparent way of indicating that one's certainty has some backing. Thus, although I might be certain that there is a God, most people would not say that they can certify that there is a God. In the 2002 EMF Risk Evaluation we developed a 0–100 "degree of certainty" scale with numerically defined categories such as: "prone to believe", "strongly believe", "virtually certain", etc. If I were to do it again I would instead call this a "willingness to certify" scale as described in Table 12.1.

In the 2002 Risk Evaluation we also went through an elaborate probability elicitation procedure providing our best-judged degree of certainty, and the lowest and highest degree of certainty that we could live with. We started with a justified

Table 12.1 A proposed "willingness to certify causality" scale adapted from the "degree of certainty" scale of the California EMF Program.

Degree of willingness to certify causality (Is Agent X at the 95th percentile exposure home or at work safe, or does Agent X increase the risk of . . . to some degree?)	Degree of willingness to certify causality on a scale of 0–100
Certify with virtual certainty that it increase the risk to some degree	>99.5
Strongly certify that it increases the risk to some degree	90–99.5
Prone to certify that it increases the risk to some degree	60–90
Close to the dividing line between certifying or not certifying that Agent X increases the risk to some degree	40–60
Prone to certify that it does not increase the risk to any degree	10–40
Strongly certify that it does not increase the risk to any degree	0.5–10
Certify with virtual certainty that it does not increase the risk to any degree	<0.5

prior degree of certainty and then elicited our certainty after having reviewed the evidence. We displayed these estimates in a graph with shaded bars.

We came up with a unique number because our decision analyst contractor Professor von Winterfeldt [8] thought that such estimates were better derived directly rather than backing into a number on the basis of using the midpoint of a broad "willingness to certify" category of the sort displayed in Table 12.1. Interestingly while many stakeholders would feel comfortable with having us declare which category best described our judgments, the display of a single number with a shaded bar around it seemed too exact to them and therefore suspect. If I had it to do again, I would simply chose a category, and insert the boundary certainties and central certainty for sensitivity analysis in the decision tree. It would come to about the same thing as using the number that resulted from the probability elicitation exercise.

It is important for decision makers and the public to recognize that this "willingness to certify" or "degree of certainty" of general causation is different from the certainties of causation talked about in legal settings. In those settings one's scientific certainty about agent X in general would have to be made relevant to the required degree of certainty of causation of *harm to a particular plaintiff* in a civil case (more likely than not) or to the required degree of certainty of *guilt of a particular miscreant* in a criminal case (beyond a reasonable doubt). To illustrate why willingness to certify general causation is different than the willingness to certify causation of a particular harm consider the following example. A scientist might be willing to certify with virtual certainty that benzene from occupational exposures will produce an extra case of leukemia among 10 000 exposed workers, but might not consider it "more likely than not" that a particular exposed worker's leukemia was due to his benzene exposure.

It is the justification of the willingness to certify a scientific degree of certainty of general causation that Bradford Hill discussed in his 1965 paper. In it he directed our attention to six "aspects" of statistical associations from the relevant epidemiological studies and three factors from other streams of evidence relevant to the issue at hand (Experimental Evidence, Plausibility and Analogy). For example, he focused on strength of the statistical association, whether the putative cause preceded the onset of disease ("temporality") and whether the alleged cause and effect relationship would theoretically imply some other observations that were then confirmed ("coherence"). In our Risk Evaluation Guidelines we reformulated his guidance in a series of questions to be answered in a graded "relative likelihood" format as demonstrated in Table 12.2.

In effect, Bradford Hill also asked certain questions about these "aspects" and gave us some rough guidance as to how some (but not all) of the possible answers to the questions ought to influence our degree of certainty that: the association reflected a truly causal effect and not the effect of (i) *unspecified* errors deriving from the conduct of the studies ("bias"), (ii) effects of *unspecified* alternative causes of the disease that tended to travel with the putative cause of interest "confounding" or (iii) the "play of chance". Bradford Hill would have assumed that before asking these questions, a good scientific reviewer would have examined the study designs and the conduct of the studies to rule out the existence of *easily identified* biases or confounding or an egregiously small study that had produced an association very easily due to chance. He did, however, caution against automatically ignoring associations that failed to meet the conventional levels of "statistical significance". "No formal tests of significance can answer those questions", Bradford Hill said. Indeed, he knew of the arbitrary origins of the conventional levels of statistical significance in the agricultural experiments of the 1920s when the great statistician R. A. Fisher decided that results dealing with crop yields and fertilizer that would have occurred only by chance in five out of 100 trials should be considered "statistically significant". This level of "significance" was then enshrined in statistical tables [9] that were used for two generations before the appearance of electronic calculators.

Bradford Hill did not formulate standardized questions related to each "aspect". Nor did he provide fixed rules to govern how the degree of certainty should be increased or decreased by the possible answers to such questions. If he had done these things he would indeed have formulated "criteria". But he did not. Also, as a sophisticated statistician he knew that no formulation of the questions could produce answers with a 100% certainty of a causal effect or a 100% certainty that an association was due only to unspecified bias, confounding and/or chance. Indeed, he wrote "None of my nine viewpoints can bring indisputable evidence for or against the cause-and-effect hypothesis".

It seems to me that Bradford Hill's "aspects" of association and Koch's "Postulates" [10] can be sorted into the two groups displayed in Table 12.3. Bradford Hill starts with the statistical association; there is a difference in the likelihood of exposure to a putative cause in cases and controls or a difference in disease rates in

Table 12.2 California EMF Program's adaptation of Bradford Hill's "aspects".

Explanations of a statistical association other than a causal one	
Chance	How likely is it that the combined association from all the studies of Factor X and disease is due to chance alone
Bias	How convinced are the reviewers that Factor X rather than a study flaw that can be *specified and demonstrated* caused this evidentiary pattern? If no specified and demonstrated bias explains it, how convinced are they that Factor X caused these associations rather than *unspecified* flaws?
Confounding	How convinced are the reviewers that these disease associations are due to Factor X rather than to another *specified and demonstrated* risk factor associated with Factor X exposure? If not due to a specified risk factor, how convinced are they that they are due to Factor X rather than to *unspecified* risk factors?
Combined effect	How convinced are the reviewers that these disease associations are due to Factor X rather than to a combined effect of chance and specified or *unspecified* sources of bias and confounders?

Attributes similar to Hill's [5] that are sometimes used by epidemiologists to evaluate the credibility of a hypothesis when no direct evidence of confounding or bias exists	
Strength of association	How likely is it that the meta-analytic association is strong enough to be causal rather than due to unspecified minor study flaws or confounders?
Consistency	Do most of the studies suggest some added risk from Factor X? How likely is it that the proportion of studies with risk ratios above or below 1.0 arose from chance alone?
Homogeneity	If a large proportion of the studies have risk ratios that are either above or below 1.0, is their magnitude similar (homogeneous) or is the size of the observed effect quite variable (heterogeneous)? How likely is homogeneity under the causal and noncausal hypothesis?
Dose–response	How clear is it that disease risk increases steadily with dose? What would be expected under causality? Under chance, bias or confounding?

(*Continued*)

Table 12.2 (Continued)

Attributes similar to Hill's [5] that are sometimes used by epidemiologists to evaluate the credibility of a hypothesis when no direct evidence of confounding or bias exists	
Coherence/visibility	How coherent is the story told by the pattern of associations within epidemiological studies? Are predicted subsidiary hypotheses confirmed? For example, if a surrogate measure shows an association, does a better measurement strengthen that association? Is the association stronger in groups where it is predicted? What would be expected under causality? Under chance, bias or confounding? How likely are the observed temporal or geographic trends if the observed magnitude of disease association with Factor X was causal versus due to bias and confounding?
Experimental evidence	How voluminous and convincing are the experimental pathology studies supporting the epidemiological evidence? What would be expected under causality, bias, chance or confounding?
Mechanistic plausibility	How voluminous and convincing is the mechanistic research on plausible biological mechanisms leading from exposure to this disease? What would be expected under causality, chance, bias or confounding? How influential are other experimental studies (both *in vivo* and *in vitro*) that speak to the ability of Factor X to produce effects at low dose?
Analogy	How good an analogy can the reviewers find with similar agents that have been shown to lead to similar diseases? What would be expected under causality, chance, bias or confounding?
Temporality	How convinced are the reviewers that Factor X exposure precedes onset of disease and that disease status did not lead to a change in exposure?
Specificity and other disease associations	How predominantly is Factor X associated with one disease or subtypes of several diseases? What would the reviewers expect under causality, chance, bias or confounding? How much is their confidence in Factor X causality for disease A influenced by their confidence that Factor X cause disease Y?

Table 12.3 Koch's Postulates and Bradford Hill's "Aspects of Association" grouped into those that pertain to a causal model and those that pertain to the quality or "pedigree" of the evidence.

Tests for causal model	Tests for pedigree of evidence
Temporality	Strength of association
Plausibility	Consistency
Expected dose–response	Homogeneity
Specificity	Koch's culturability
Coherence	Experimentation
Analogy	(a) Causal (Koch's animal model)
	(b) Preventive

exposed and unexposed cohorts. Two of Koch's postulates if satisfied would imply that a statistical association would have to be present (the high likelihood of microbes in those with the disease syndrome and a low likelihood of microbes in those without the disease syndrome) although they typically are presented in an all or nothing fashion. The nine Bradford Hill "aspects" and the remaining Koch's "postulates" are displayed in Table 12.3.

In the left hand column are those that could be thought of as characteristic elements of a causal diagram of the generic sort shown in Fig. 12.1.

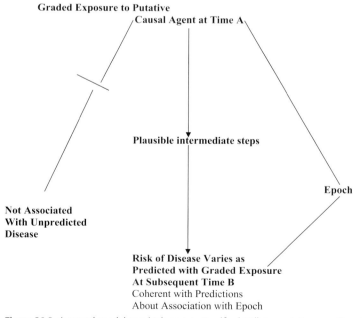

Figure 12.1 A causal model to which certain Bradford Hill "aspects" are pertinent.

If present these would increase ones willingness to certify causality. Temporality relates to the fact that cause precedes effect in time. Graded exposure produces a graded effect ("dose–response"). Information from other sources on mechanism suggests plausible intermediate steps (plausibility). If a putative exposure is known to be associated with another factor, say a higher prevalence of exposure in one "epoch" of time than in another, then one would predict that the disease too would be more prevalent in that epoch. If this is confirmed, the story is "coherent".

The exposure is not associated with unexpected diseases (specificity). The requirement of specificity grew from a belief of Conn and Koch [10] that species of microbes were physiologically unique and would therefore cause unique diseases. We have since learned that this is not always the case, e.g. viral and bacterial causes of sore throat are hard to distinguish clinically, and the agent of Syphilis *Treponema pallidum* can cause a variety of quite distinct syndromes. The smoking habit is associated with a variety of diseases.

The other "aspects" of an association mentioned by Bradford Hill and some of Koch's postulates relate more to what Silvio Funtowicz [11] has called the "pedigree" of the evidence, i.e. they relate to the quality of the evidence more than they relate to the proposed causal model. Thus, regardless of what the model would predict, the stronger the association, the less likely it is due to unspecified confounding or bias. Similarly if one can produce or prevent the disease experimentally by applying the putative agent or withdrawing it (an impressive methodological pedigree indeed), one's certainty of causality increases.

12.4
The Asymmetry of Some "Rule In" Tests

Bradford Hill warned us at several points in his 1965 article that it is really quite rare, even in studies of well recognized causal agents, to get a "yes" answer to the question "Is the observed association a lot stronger than the usual strength of associations produced by confounding or bias?". With a few notable exceptions such as the smoking habit, some virulent pathogens, etc., most causal agents are present in amounts that produce modest associations. It should be remembered, however, that studies of noncausal agents would almost never yield a "yes" answer to this question. Since a "yes" answer to this question is more likely if the association reflects true causality than if the association were due to bias, confounding or chance, a "yes" answer justifies a substantial increase in one's degree of certainty that this was indeed a causal association. On the other hand, how much ought one's certainty in causality be influenced by an answer: "No the association is NOT much larger than the associations often produced by bias, confounding and chance?". This answer is commonly found both in studies of causal agents and in studies of noncausal agents. Thus, a "no" answer justifies little or no decrease in one's certainty about causality. The impact on certainty from a "yes" or a "no" answer to this Bradford Hill question is thus asymmetrical. It can help to rule in causality, but it is not very good at ruling it out. This is why Bradford Hill wrote: "We must not be too ready to dismiss a cause-and-effect hypothesis merely on the grounds that the observed association appears to be slight".

This kind of asymmetry is recognized in the law as well. A witness who is known to be well disposed to an accused felon and who has no self-interest in cooperating with the prosecution will be readily believed if they provides incriminating evidence against his friend, but their testimony will be taken with a grain of salt if they provides exonerating evidence.

12.5
Toulmin's Argument Anatomy and Bayes' Theorem as a Universal Warrant

The theory of argument propounded by Stephen Toulmin in his classical 1958 book *Uses of Argument* [7] can be useful here. Toulmin suggests that all nondeductive argument (the kind of argument used in science, policy, ethics and aesthetics) has a similar structure. They all base their "claims" on factual "grounds" and a general inferential rule that "warrants" a claim based on those grounds. There must be "backing" for this inferential rule and there may be "rebuttals."

Figure 12.2 diagrams a medical diagnostic argument. The claim is made that one should increase one's certainty that a patient with chest pain has myocardial infarction (MI) as opposed to some other cause. The general inferential rule is guided by Bayes' theorem, i.e. that one is warranted to increase ones degree of certainty that a chest pain patient has an MI in proportion to the product of the relative likelihood conveyed by having a particular blood enzyme level and the pre-test probability of MI in one's clinic population [12]:

$$\text{Post-test odds} = \text{pre-test odds} \times \text{relative likelihood of test results}$$

The backing for Bayes' theorem is the axioms of probability theory. The backing for the warrant regarding the relative likelihood conveyed by blood enzyme tests results comes from previous studies comparing chest pain patients with MI to all other chest pain patients with respect to their blood enzyme distributions. The grounds of the argument include the prevalence of MI in this clinic's usual population, which gives us the "pre-test" certainty, as well as the fact of chest pain and the particular blood enzyme level. The rebuttal might be that the patient was or was not kicked in the chest by a mule.

Let us now apply Toulmin's argument scheme to see what can be claimed on the basis of Bradford Hill's "strength of association test".

Test: How does the observed strength of association compare to the associations that can occur from unspecified bias and or confounding?

Warrants for possible result patterns:

(a) If the observed association is much larger than associations previously known to be produced by bias and confounding, this result pattern is more likely among a hypothetical population of true causes than among a similar population of noncauses. Thus we generate a "heuristic relative likelihood" and we judge it

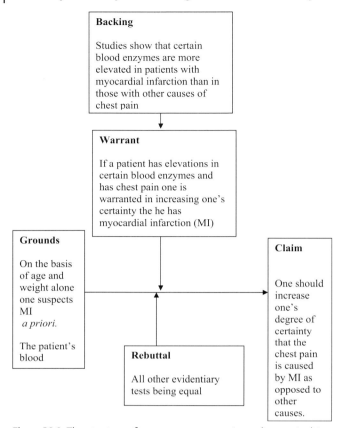

Figure 12.2 The structure of an argument supporting a diagnostic claim.

to probably be much larger than 1.00. According to Bayes' theorem one is warranted to substantially boost one's certainty above the "pre-test" certainty.

(b) If the observed association is about the same size as that produced by bias and confounding, this would be seen both in true causes and noncauses so the heuristic relative likelihood is close to or slightly less than 1.0. Following Bayes' theorem we are warranted to move below "our pre-test" certainty only slightly.

(c) If there is no association, that would be a pattern unlikely among true causal agents and likely among noncausal agents. The heuristic relative likelihood is thus less than 1.00 and according to Bayes' theorem one is warranted to move one's post-test certainty below one's "pre-test" certainty that the agent causes adverse effects.

Notice the asymmetry in warrant (a) and warrant (b). This can happen when a test has an extremely low nonspecificity in the face of a low sensitivity. For example, if a test had a 10% sensitivity and a 1% nonspecificity, a positive result would convey a likelihood ratio of 10%/1% = 10, while a negative result would convey a likelihood ratio of 90%/99% = 0.91. Thus, a positive result boosts us far above our pre-test certainty (whatever it was), while a negative result leaves us pretty much where we were to begin with.

A similar approach could be taken to Koch's question about finding an animal model in which one could introduce a pure culture of the agent and produce the disease syndrome, with the agent teeming within. It makes much more sense when considered as an insensitive but specific test than as an absolute requirement. The probabilistic version is presented below.

Test: What happens when four "animals" like Professor Max von Pettenkofer and his three assistants, Stricker, Metchnikoff and Ferran, drink 1 cm^3 of bouillon culture teeming with *Vibrio cholera* taken 4 days before from the diarrhea of a man who died of cholera?

Warrants for possible result patterns:

(a) If all four develop a full case of cholera or dies from it and had the vibrio in their stool, this would be unlikely if the vibrio did not cause the disease and likely if it did, so the heuristic relative likelihood is much greater than 1.0 and one is warranted to increase one's posterior degree of certainty of vibrio causality substantially (Karl Popper not withstanding).

(b) At least one but not all subjects develop full-blown cholera symptoms and have vibrio in their stools. This result pattern is extremely unlikely if the vibrio is incapable of causing the cholera syndrome and quite likely if the vibrio can cause a 25% or more attack rate.

(c) If they do not develop cholera yet have vibrio in their stools (as was indeed the case [13]) this would be less likely if the vibrio causes a 100% attack rate of cholera and more likely if it causes a low attack rate or is not a cause at all. So one is warranted to move to a post-test certainty of a high attack rate cause that is lower than the pre-test certainty and a higher post-test certainty of either a low attack rate agent or an agent that cannot cause cholera symptoms at all except after passage through the soil.

Pettenkofer seemed to have assumed that if the vibrio could directly cause severe cholera symptoms without passage and transformation in the soil, the attack rate would be so high that an experiment with only four subjects would be sufficient to demonstrate the effect and the likelihood of four asymptomatic or mildly ill patients was virtually zero under the causal hypothesis, while it was to be expected under the null hypothesis [14]. Koch probably expected the same thing. The language of his postulates is very all or nothing.

Both Pettenkofer and Koch were somewhat vague as to what claims could be warranted by the possible results of this experiment prior to Pettenkofer's embarking on it. Afterward, Pettenkofer claimed that he had definitively falsified Koch's theory (Popper might have agreed), and Koch and his allies claimed that Pettenkofer and his three assistants, Stricker, Metchnikoff and Ferran, who repeated the experiment with approximately the same result must have been resistant (i.e. the likelihood of full-blown cholera under the causal hypothesis was not 100% after all as required under Koch's animal model postulate) [10]. However, Koch once again framed the survival of his Munich antagonists in a deterministic way. They belonged to a class of humans who had a zero risk of developing symptoms after ingesting the vibrio. The point here is that Koch and Pettenkofer, like people today, would benefit by making their "warrants' explicit (as we have done above) prior to doing a study, rather than modifying them after the fact to fit their hypotheses. Indeed many scientific arguments arise because of differing unspoken assumptions about the proper warrant to use. By making their warrants explicit, opposing theoretical camps would be forced to examine their assumptions and the scientific "backing" that supports them. For example, Koch and Pettenkofer might have been forced to consider what percent of the population in places where previous cholera epidemics had hit, and in populations where cholera had always been unknown, would develop the full-blown disease if they were submitted to Pettenkofer's type of experimental exposure. Was the fact that not everyone developed cholera during an epidemic due to differing exposure, differing innate and acquired resistance or both?

This asymmetry in the impact of "yes" and "no" answers is true for the remaining Bradford Hill questions and for Koch's Postulates as well. The Bradford Hill aspects of "temporality" and "coherence" may be an exception.

With regard to "temporality" it is usually, but not always, easy to demonstrate that the putative cause occurred prior to the induction of the disease process and evidence for this boosts our certainty in causality. Good evidence that this is *not* so justifies us in lowering our certainty of causation to zero.

12.6
Special Importance of Coherence

The "aspect" of "coherence" received only a brief comment and vague definition from Bradford Hill, and is often given short shrift in causal discussions. Rothman [15] rightly emphasizes in his discussions on inductive reasoning that if a change in exposure to agent X truly causes a change in the risk of disease Y there are usually other facts about the state of the world that could be deduced to be the case. Within the limits of the evidentiary tests to confirm these predictions, the degree of observational confirmation of the deductions should increase our degree of certainty that there is indeed a causal association at work. The lack of confirmation of the deductions should lower our degree of certainty. Although I am not aware of any empirical backing for this assertion, I present a thought experiment as backing. Think of the population of true causes that have associations with other variables. One predicts that these other variables will be associated with disease and the likelihood that these predictions are confirmed is high. Then consider a population of agents that because of confounding seem to be associated with a disease, but are not themselves causes. We know of associations between these noncauses and other variables, and predict that the disease will be associated with these other variables. However, will the true hidden cause also have direct associations with these other variables or will it only be associated indirectly through the noncause? Some of the time there will be no direct associations between the hidden cause and the variables associated with the noncause. It seems reasonable then to suppose that the likelihood of confirming the coherence predictions is lower in a population of noncauses than in a population of causes. One could examine the epidemiological literature in a meta-analytic mode to confirm that this heuristic insight is justified. Bradford Hill and others seem to accept it as given.

Bates [16], for example, examined the hospital admissions for respiratory disease, respiratory symptoms and respiratory function tests in polluted and nonpolluted areas because the higher respiratory mortality rates in more polluted areas suggested that one ought to find these other differences as well. When this prediction was confirmed, this increased his belief that the pollution caused the deaths. If his predictions had failed his certainty that the pollution/respiratory death association was causal in nature would have been pushed downward. Thus, "coherence" reasoning can either increase or decrease our degree of certainty in causation and should more frequently be employed

12.7
Plausibility, Experimentation and Analogy

Epidemiology, of course, is not the only relevant stream of evidence. For example, in the case of potential health effects from EMFs, we had to consider streams of evidence from biophysical calculations, chemistry, molecular biology, cellular biology, animal physiological experiments, animal toxicology, animal observation, human physiological experimentation and epidemiology. It has been my observation that scientists from each of the fields tend to pay attention primarily to the concepts and evidence that derives from their particular field. If that evidence is convincing, they are prone to accept results from other streams of evidence more readily. In particular, scientists from experimental fields look with great doubt on nonexperimental epidemiological results. So as a matter of sociological fact, a claim that rests on factual "grounds" from many streams of evidence will only get universal approval if the evidence from most of the streams points in the same direction. Indeed a heuristic that is used for deciding if an agent is hazardous is that a "good story" can be told about how the agent acts at each level of physical, chemical and biological organization. Figure 12.3 illustrates this heuristic using dominos.

Here, an EMF causes a domino to fall that represents a physical induction mechanism. The falling of this domino causes a domino representing a chemical event to fall. The chemical event causes yet another domino to fall representing a molecular biology event. This in turn hits another domino representing a cellular biology event. This continues toward whole-organism physiological events, some of which produce no other effect and are "dead ends". In another branch, however, the chain of causation continues onward causing disease in certain strains of animals,

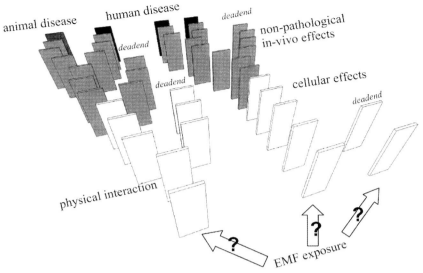

Figure 12.3 Dominos representing steps in a mechanism linking the exposure to EMFs with cellular effects, nonpathological *in vivo* effects and disease.

but not in humans, while in another branch, the chain can cause disease only in humans or both in animals and humans.

Clearly there is no single experiment in which all these events at different levels of organization are observed to happen in sequence. Rather, the "story" is pieced together from many experiments over time. Under this heuristic, the more steps in the story that one can fill in, the more one believes that the link between EMF and disease (between the first and last domino) is causal in nature. However, how does a risk evaluator decide how to assign a score to "how good the story is?".

From a Bayesian point of view each stream of evidence as to whether a particular "domino" has fallen in the above-depicted model can be looked at as an independent (or intercorrelated) test. Even before looking at the actual pattern of test results from that stream one can assess its inherent sensitivity and specificity, and therefore its ability to convey useful relative likelihoods from apparently exonerating or incriminating results. This usefulness of a stream of evidence will vary from agent to agent. There are some "tests" which are capable of both exonerating and incriminating an agent. Others which are primarily exonerators or primarily incriminators and some tests that can neither incriminate nor exonerate. If the tests embodied by each stream were completely independent, one would simply multiply the relative likelihoods conveyed by the observed results from each stream of evidence. For example, streams that for this agent were prone to falsely exonerate, but capable of incriminating, would convey relative likelihoods near 1.00 for exonerating evidence, but relative likelihoods far above 1.00 for incriminating results.

Table 12.4 Example of a summary table of qualitative relative likelihoods from the California EMF Risk Evaluation document [3].

	Summary Table for Childhood Brain Cancer		
	How likely is this attribute of the evidence under:		How much and in what direction does this attribute change confidence?
Attribute of the Evidence	"No-effect" hypothesis	Causal hypothesis	
Chance is credible explanation.	likely		chance has not been ruled out
Upward bias not suggested for body of evidence	possible	possible	none
Confounding unlikely	possible	possible	none
Combined, chance, bias, confounding	likely	possible	chance has not been ruled out
Strength of association does not exceed possible confounding or bias	possible	less possible	no impact or slight decrease
Not consistently above the null	possible	less possible	no impact or slight decrease
Homogeneity lacking between size of effects in few positive studies	possible	less possible	no impact or slight decrease
Dose–response not clear in studies	possible	less possible	no impact or slight decrease
Coherence/visibility: temporal trends would not reflect these near-null effects	possible	possible	none
Experimental evidence for brain tumors is null	possible	less possible	no impact or slight decrease

In the California EMF Risk Evaluation we presented a table similar to that in Table 12.4 (for brain cancer) for each of the 13 diseases that we examined and indicated for each of Bradford Hill's questions how likely the pattern of evidence was under (i) the null hypothesis of chance, unspecified bias or confounding, or (ii) the causal hypothesis. This led us to a series of qualitative relative likelihoods.

However, this left us with the final leap from this series of qualitative relative likelihoods to a willingness to certify general causality. This was not unlike the leap that a physician has to make to a willingness to certify the existence of a particular diagnosis after considering the results from all the tests they have run, and the results of the history and physical they have conducted. In our case we asked ourselves which willingness-to-certify category most described our assessment after examining this table. This had the advantage of forcing us to systematically consider the complete evidentiary pattern. Then we provided verbal accounts explaining what most swayed our opinion.

Our approach differed from that used by the IARC in a number of ways:

(a) Our final judgments used a system that was more finely graduated and used specified boundaries of the certainty categories so that they could be applied in decision analysis. The IARC "possibly carcinogenic" category could mean "not impossible, but highly improbable" or "just shy of probable". Indeed, it includes agents like coffee, which I doubt is carcinogenic, and fiber glass, which is very likely to be.

(b) Using qualitative relative likelihood tables and verbal justifications makes the argument warranting the claimed willingness to certify much more transparent.

(c) IARC demands that weight be given to toxicological results even if there is *a priori* evidence that in the particular agent, the "well-conducted toxicological test" is prone to falsely exonerate so that null results would not convey relative likelihoods much different from 1.00.

(d) The IARC system of classification is really a classification of evidence results and NOT a classification of rational probability of carcinogenicity. This is true even though some of the categories have such labels as "not carcinogenic", "possible", "probable" and "definite". Another IARC category is more transparently labeled "inadequate evidence". It would be more transparent if IARC simply used labels descriptive of result patterns such as:
  (I) Strongly incriminating toxicological and epidemiological patterns of evidence
  (II) Moderately incriminating toxicological and strongly incriminating epidemiological patterns of evidence
  (III) Exonerating toxicological evidence and moderately incriminating epidemiology
  (IV) Insufficient volume of evidence
  (V) Conflicting or poor quality toxicological and or epidemiological evidence
  (VI) Strongly exonerating toxicological and epidemiological evidence of high volume and quality

  If they did this it would become clear that professional judgment would be needed to define how one was to use words such as "moderately" "insufficient volume", "poor quality" and "strongly".

(e) IARC's unexplained leap from results categories such as "strong toxicological incrimination and strong epidemiological incrimination" to rational probability claims such as "willing to certify carcinogenicity with virtual certainty" is almost tautological because that is how we have agreed to define carcinogens. One could explicitly justify this convention if one considered the likelihood of such a results pattern in a population of noncarcinogens (very, very, very low likelihood) and among a population of carcinogens (not so likely, but relatively much more likely than among noncarcinogens). The relative likelihood of this pattern is thus very far above 1.00 and ought to boost one's post-test betting odds away from the pre-test betting odds. However, the IARC ignores the pre-test betting odds, and declares all agents with strongly incriminating toxicological and epidemiologic evidence as definite carcinogens.

Also, consider the result pattern category of "'null' toxicology + incriminating epidemiological associations not much above the resolution power of the studies". Here, our imagination for likelihoods among carcinogens and noncarcinogens is much less clear. Yet, the IARC is iron clad as to what category agents with such results patterns should fall into and how suspicious we should be about it. However, that practice is problematical. In fact, depending on the agent at hand, the toxicological test might differ as to its sensitivity and specificity. There should not be the same one-to-one mapping between result pattern and willingness to certify causality as there could be at the top of the scale. Indeed, even at the bottom of the result pattern scale, where toxicology to date and epidemiology to date are strongly exonerating, I cannot justify the same degree of willingness to certify safety (a new animal model may be found, a longer follow-up period may reveal epidemiological findings missed before) as I could in certifying carcinogenicity on the basis of strongly incriminating toxicology and epidemiology.

The assumption underlying the IARC and other current risk assessment paradigms is that there is the possibility of a decontextualized invariant way to produce degrees of certification on the basis of predefined categories of evidentiary result patterns, without considering the sensitivity and specificity for this particular agent. I believe that there are situations in which this optimistic assumption can be misleading. The qualitative Bayesian warrant which considers pre-test odds and relative likelihood of evidentiary result patterns is conceptually a more valid way to proceed.

12.8
Causal Arguments Can Become More Transparent but Will Always Involve Judgment

Judging by his article, Bradford Hill probably expected that the listed aspects of the observed statistical association would be used as paragraph headings for a thoughtful examination of systematically gathered facts and a thoughtful application of empirically backed general rules of inference to justify claims for increased or decreased degrees of certainty of causation. He was not advocating a "yes/no" check list approach for evaluating each successive aspect nor did he advocate a simple *a priori* weighting of separate scores to solve the thorny problem of how much the insights derived from the answer to each separate question should contribute to the final degree of certainty. He did not assume that causality could be diagnosed with a checklist anymore than one would have assumed that a physician would make a diagnosis by checklist or a judge would derive a judgment by checklist. All these procedures require professional judgment. A completely quantitative Bayesian approach of the sort proposed by McColl et al. [17] or by Lindley [18] would require assigning many quantitative parameters to a complex Bayesian Net model that would mathematically combine the subjectively assigned parameters to produce a posterior willingness to certify causality. This has never to my knowledge actually been done while including all streams of evidence. How experts such as physicians combine streams of evidence to make judgments about causality has been of great practical interest. As pointed out by Shortliffe [19] there have been two general approaches – one is to interview or observe experts obtain the warrants that they use, the other is to develop warrants using the backing of actual data on the relative likelihood conveyed by different patterns of results. An example of the latter was Dedombal's [20] work on diagnosing the cause of acute abdominal pain. According to Shortliffe [19] neither approach has allowed computers to replace professional judgment in medical diagnosis. The same will therefore be true for risk evaluation. Toulmin, in his book *Return to Reason* [21], points out that the hope for a universal Euclidean deductively certain style of reasoning of the sort possible in some domains of physics grew out of a craving for objective certainty after the theological blood baths of the 30 Years War of the 17th century. We are learning, as Weed [22] has reminded us, that there are domains like the law, medical diagnosis and risk assessment were this particular kind of reasoning cannot be hoped for. Indeed, Gigerenzer [23] has argued that heuristics that are even faster and more frugal than the qualitative Bayesian approach we used in California may be "good enough". Nonetheless, clear rules for accepting evidence and a structured argumentation to warrant claims for degrees of certainty *can* increase reliability and transparency.

It should be noted that the rules that govern the types of study that are admitted for consideration and the general inferential rules that "warrant" certain degrees of certainty can be chosen in ways that influence the balance between falsely exonerating and falsely incriminating an agent as hazardous. The public interest is served by allowing scientific reviewers to consider and weigh a wide range of evidence, and to use rules of inference that give more weight to avoiding false exoneration than might be the case in the evaluations often done to guide research strategy.

It should also be noted that the kind of structured, transparent and formalized evaluation described above, aimed as it is to certifying a degree of certainty at one point in time to assist some societal decision is *not* a common procedure in the usual course of doing science. Although familiar scientific concepts and skills contribute to and indeed are crucial to the process, it requires additional skills and experience to do it properly some of which resemble the skills of legal argument.

Once the rules of evidence and inference have been agreed upon at the outset, it is incumbent on the scientists involved to proceed impartially before the court of science, in regulatory settings or before the bar. As Bradford Hill [5] said:

> The evidence is there to be judged on its merits and the judgment (in that sense) should be utterly independent of what hangs on it – or who hangs because of it.

References

1. Neutra, R.R. and DelPizzo, V. (2002) Transparent democratic foresight strategies in the California EMF Program, *Public Health Reports*, **117**, 553–563.
2. IARC (2001) *Static and Extremely Low Electric and Magnetic Fields*, International Agency for Research on Cancer, Lyon.
3. California Department of Health Services (1999) *Electric and Magnetic Field Risk Evaluation Guidelines*, California Department of Health Services, Oakland CA [http://www.dhs.ca.gov/ehib/emf/newsactostyle.pdf] [Retrieved: 16.02.2007].
4. Neutra, R., DelPizzo, V. and Lee, G.M. (2002) *An Evaluation of the Possible Risks from Electric and Magnetic Fields (EMFs) from Power Lines, Internal Wiring, Electrical Occupations and Appliances*, California Department of Health Services, Oakland, CA [http://www.dhs.ca.gov/ehib/emf/RiskEvaluation/riskeval.html] [Retrieved: 16.02.2007].
5. Hill, A.B. (1965) The environment and disease: association or causation? *Proceedings of the Royal Society of Medicine*, **58**, 295–300.
6. Evans, A.S. (1976) Causation and disease – Henle–Koch postulates revisited, *Yale Journal of Biology and Medicine* **49**, 175–195.
7. Toulmin, S. (1958) *The Uses of Argument*, Cambridge University Press, Cambridge.
8. von Winterfeldt, D., Eppel, T., Adams, J., Neutra, R. and DelPizzo, V. (2004) Managing potential health risks from electric powerlines: a decision analysis caught in controversy, *Risk Analysis*, **24**, 1487–1502.
9. Fisher, R. and Yates, F. (1963) *Statistical Tables for Biological, Agricultural and Medical Research*, 6th edn, Oliver & Boyd, Edinburgh.
10. Carter, K. (1985) Koch's postulates in relation to the work of Jacob Henle and Edwin Klebs, *Medical History*, **29**, 353–374.
11. Funtowicz, S. and Ravetz, J. (1990) *Uncertainty and Quality in Science for Policy*, Kluwer, Dordrecht.
12. Sackett, D., Haynes, R., Guyatt, G. and Tugwell, P. (1991) *Clinical Epidemiology*, Lippincott Williams & Wilkins, Philadelphia, PA.
13. Brock, T. (1988) *Koch, Robert: A Life in Medicine*, Science Technology Publishers, Madison, WI.
14. Hume, E. (1927) *Max von Pettenkofer: His Theory of the Etiology of Cholera*, Paul B. Hoeber, New York.
15. Rothman, K. (2000) *Epidemiology: An Introduction*, Oxford Press, New York.
16. Bates, D.V. (1992) Health indexes of the adverse effects of air pollution – the question of coherence, *Environmental Research*, **59**, 336–349.
17. McColl, R.M., Husted, J. and G.M., P. (1996) *A Framework for Assessing and Combining Evidence for the Carcinogenicity of Environmental Agents: A Prototype System*, Institute for Risk Research, University of Waterloo, Waterloo.
18. Lindley, D.V. (2000) The philosophy of statistics, *Journal of the Royal Statistical Society Series D: The Statistician*, **49**, 293–319.
19. Shortliffe, E.H., Perreault, L., Wiederhold, G. and Fagan, L.M. (eds) (2001) *Medical Informatics: Computer Applications in Health Care and Biomedicine*, Springer, New York.
20. Dedombal, F.T., McCann, A.P., Leaper, D.J., Stanilan, J.R. and Horrocks, J.C. (1972) Computer-aided diagnosis of acute abdominal pain, *British Medical Journal*, **2**, 9–13.
21. Toulmin, S. (2001) *Return to Reason*, Harvard University Press, Cambridge. MA.
22. Weed, D.L. (1997) Underdetermination and incommensurability in contemporary epidemiology, *Kennedy Institute of Ethics Journal*, **7**, 107–127.
23. Gigerenzer, G. and Todd, P. (1999) *Simple Heuristics That Make Us Smart*, Oxford University Press, New York.

13
Evidence Maps – A Tool for Summarizing and Communicating Evidence in Risk Assessment

Holger Schütz, Peter M. Wiedemann, and Albena Spangenberg

13.1
Introduction

Evidence maps are an approach to evidence characterization that aims at improving the transparency and reasonableness of reporting scientific evidence regarding the existence of a hazard or a risk. They are designed to depict the reasons leading experts to their conclusions about a (potential) hazard or risk. Evidence maps provide a graphical representation of the arguments that speak for or against the existence of a causal relationship between exposure to a (potentially) hazardous substance or condition and the biological effects that are considered, as well as the conclusions that are drawn and the remaining uncertainties. Although the logical structure of evidence maps is independent of a particular problem – in fact, they should be applicable to any assessment that has to be substantiated – the practical value of evidence maps depends on the quality of their input. In order to facilitate an evaluation of this quality by the reader, evidence maps should be accompanied by a description of the process through which they have been generated.

In the following, we will therefore not only outline the evidence maps approach, but we will also give an example of how to organize the process of generating input for building evidence maps.

13.2
Evidence Maps Approach

13.2.1
Background

The evidence maps approach was developed within a project that started in September 2003 and was completed in April 2005 (the project was supported by T-Mobile, Germany). It aimed at conducting a transparent process of risk evaluation with

The Role of Evidence in Risk Characterization: Making Sense of Conflicting Data.
Edited by Peter M. Wiedemann and Holger Schütz
Copyright © 2008 WILEY-VCH Verlag GmbH & Co. KGaA, Weinheim
ISBN: 978-3-527-32048-6

respect to mobile telephony and health. The electromagnetic fields (EMFs) considered cover the frequency range from 900 to 2000 MHz. Six topic areas which were at the center of the scientific – but also public – debate were selected for evaluation:

- Genotoxic effects
- Laboratory animal experimental studies of cancer
- Epidemiological cancer studies
- Effects on the central nervous system in the awake and sleep states
- Impairments of well-being
- Effects on the blood–brain barrier

Each area was assessed by two experts from the respective field. Their task was to select and evaluate the most important scientific studies published between the years 2000 and 2004 for the respective topic area. In order to ensure a highest degree of consistency – and consequently comparability – of the presentation of the state of scientific knowledge, we suggested a unified structure for the expert opinion reports that was adopted by all the experts. (Our suggestion for the structured presentation of the expert opinion reports was guided by the Cochrane Review, see http://www.cochrane.dk/cochrane/handbook/3_1_rationale_for_protocols.htm, http://www.cochrane.dk/cochrane/handbook/appendix_2a_guide_to_the_format_of_a_cochrane_review.htm and http://www.cochrane.dk/cochrane/handbook/hbook.htm). Table 13.1 lists the specific points addressed by the expert opinion reports.

For each topic area, both experts were asked to closely coordinate their efforts on the first two work steps. In doing so, the expert opinion reports would pertain to the same endpoints and be based on the evaluation of the same set of primary studies. Originally, two separate expert opinion reports were called for; however, in several cases the experts realized that there is a significant overlap – even for the evidence evaluation. In these cases they decided to develop a joint expert opinion and

Table 13.1 Suggested structure of expert opinion reports.

Aim of the expert opinion report
- Characterization of the topic area (especially in regards to the relevance of the findings from this topic area for the evaluation of potential health risks)
- Selected endpoints and rationale behind their selection

Selection of the studies to be considered from the period 2000 to 2004
- Criteria for the selection of the considered studies (if necessary, also mention of the selected field strengths, frequency range and signal shape)
- Search strategies for the selection of primary research studies (personal bibliographic lists of references/databases; Medline, etc.)
- Information on the quality of method for each study

Presentation of the state of scientific knowledge
- Discussion of the findings and method of the studies for each individual endpoint
- Evaluation of the scientific weight-of-evidence for the individual endpoints

Overall evaluation for the topic area
- Summarizing evaluation of the scientific evidence for the topic area

List of the considered studies and the references used

individually point out the possible evaluation differences within a joint report (this was the case for the topic areas "genotoxic effects", "experimental laboratory animal cancer studies", "epidemiological cancer studies" and "blood–brain barrier", while for "CNS/sleep" and "impairment of well-being" separate expert opinion reports were prepared).

An essential criterion for the selection of the experts was their involvement in EMF research and expertise in the respective area, as documented by publications in recognized specialty journals. This was to ensure that the experts would have both the technical background as well as the methodological knowledge for preparing an expert opinion report. Knowledge gained from performing one's own experimental studies as well as empirical expertise with the investigation methods is crucial for the critical evaluation of the scientific evidence presented in the primary studies on possible health effects from radiofrequency (RF) EMFs.

The experts or expert groups provided an opinion report which described the procedure and results of their review of the scientific literature, according to the structure presented in Table 13.1. For each topic area, the expert opinion reports were critically discussed at workshops with advisory expert panelists. Selection of advisory experts was based on their scientific expertise in the respective topic area. However, it was not required that the scientific research of the advisory experts focuses specifically on the EMF field (although this was for the most part the case). Rather, the selection depended on their ability – for the respective topic area – to critically review the expert opinion reports with respect to the theoretical underpinning and methodological rigor. The selection of the experts and the advisory expert panelists was performed by the authors. An attempt was made to find experts that are representative of the whole spectrum of scientific opinion for each topic area. Altogether, 25 leading scientists from Germany and Switzerland were involved in this project.

These expert reports provided the input for the evidence maps which we constructed for each topic area. The evidence maps were presented at a final workshop and discussed with all experts involved in the project.

13.2.2
Structure of Evidence Maps

The theoretical basis of evidence maps approach is 2-fold. First, Toulmin's argumentation theory [1, 2], which includes four main components: data, warrant, backing and claim. In our case *data* refers to the scientific evidence provided by the available scientific studies on the respective topic. The *claim* refers to the conclusion whether a causal relationship exists between exposure to an agent and an adverse health effect. The *warrant* consists of "pro" and "con" arguments which link the data to the claim. These pro and con arguments may be supported or attenuated by additional arguments (*backing*). Conceptualizing the warrant as pro and con arguments introduces the second theoretical perspective: the weight-of-evidence (WOE) approach. Although there is no standard methodology for performing a WEO evaluation [3–5], the basic idea behind it is obvious: all scientifically sound evidence should be included in a risk assessment according to its methodological rigor, and positive and negative evidence should be weighed against each other. The following citation from the US Environmental Protection Agency (EPA) illustrates this ([6], pp. 71–72):

> Risk assessment involves consideration of the weight of evidence provided by all available scientific data. In other words, "weight of evidence evaluation is a collective evaluation of all pertinent information so that the full impact of biological plausibility and coherence is adequately considered" . . . .
> Judgment on the weight of evidence involves consideration of the quality and adequacy of data and consistency of responses induced by the stressor. The weight-of-evidence judgment requires combined input of relevant disciplines: toxicology, biology, chemistry, epidemiology, statistics, etc. Initial views of the database may change significantly when other data are brought into consideration. For example, the impact of a positive animal carcinogenicity study may be diminished by high-quality negative studies; or a weak association in human epidemiologic studies may be bolstered by consideration of other key data from animal or other assays. Generally, no single study, whether positive or negative, drives the overall weight-of-evidence judgment. And study findings are not scored by any mathematical algorithm; rather, they are based on professional scientific judgment.

By combining Toulmin's argumentation theory with the WOE approach, evidence maps try to depict the underlying "reasoning" leading experts to their conclusions about a (potential) hazard. To this end, the core elements of risk assessment are presented in a graphical manner: (i) the evidence basis, or the data, (i.e. the number and the quality of available scientific studies), (ii) the pro and con arguments, the warrants, with their respective supporting and attenuating arguments, and (iii) the conclusions, or claim, about the existence of a hazard with (iv) the remaining uncertainties. Figure 13.1 shows the template for an evidence map.

In addition, the evidence map also provides information about the knowledge base, i.e. the number of studies conducted with respect to the selected endpoint, as well as the proportion of the studies that comply with the required methodological quality standards.

In our view, the overall structure of the evidence map outlined in Figure 13.1 enhances the understanding of the expert opinions about risks and supports their critical evaluation. It discloses both the available evidence as well as the remaining uncertainty. Furthermore, it gives a fair picture of both sides of the available evidence – the pro-risk as well as the contra-risk arguments.

13.2.3
Constructing an Evidence Map: Cancer Epidemiology

In the evaluation of potential risks associated with RF EMFs from mobile telephony, epidemiology plays an important role. Epidemiology aims at the identification and characterization of causes responsible for diseases in humans. However, complex exposure conditions as well as multifactorial cause–effect relationships can make the proof of causation for environmental health-related diseases very difficult.

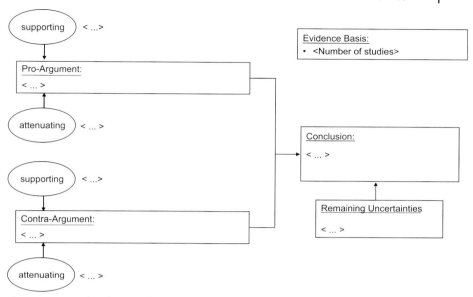

Figure 13.1 Template for an evidence map.

Nevertheless, epidemiological studies have the advantage of avoiding the uncertainties that are inherent to extrapolation of animal models to humans, because they directly test "in humans".

Out of 122 studies identified for the period from 2000 to April 2004, the experts in our project selected 13 scientific studies as acceptable for risk evaluation [7]. All studies referred to exposures to EMFs from mobile phones, because during this period no scientific studies were published on the potential health risks associated with mobile phone base stations.

Figure 13.2 shows the evidence map which summarizes the results of the experts' evaluation. From the map, it becomes obvious, that there is both a central pro and a con argument. Both arguments, however, are attenuated by methodological deficits of the studies.

While an evidence map can depict the structure and the main arguments of the experts' evidence characterization, it cannot include a detailed evaluation of the scientific studies reviewed by the experts. Therefore, an evidence map must be supplemented by a written summary of these evaluations. The following example from the area of cancer epidemiology can illustrate the procedure.

In five studies, noticeable results are found pointing to a relationship between exposure to mobile telephony fields and cancer. One of those is the study by de Roos et al. [8] on neuroblastoma. For the fathers of toddlers with neuroblastoma, no increased risk is found; in the mothers an increased odds ratio of 1.8 [95% confidence interval (CI) 0.5–6.0] from occupational exposure to mobile telephony in the last 2 years preceding the pregnancy is determined. A methodological deficit of this study is the imprecise estimation of the effect based on the low number of exposed study participants. The data collection on mobile phone utilization was conducted with

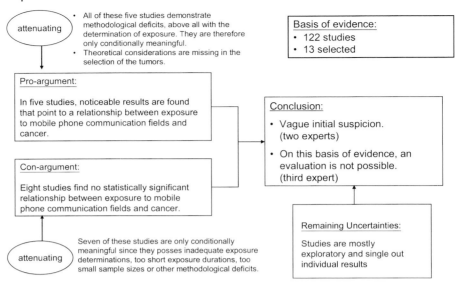

Figure 13.2 Evidence map for cancer epidemiology.

little detail and referred merely to the workplace. Nevertheless, the noticeable maternal result cannot be simply explained as a result of confounding (private mobile telephone usage), because then the confounder *per se* would be responsible for the health risk. Despite the study's serious methodological deficits, the experts view the results as "noticeable findings" and "first indications of a relationship between mobile telephony exposure and risk of neuroblastoma". To no small extent, this is because there was no prior knowledge present of a relationship between mobile telephony and neuroblastoma, and because even in the future a study of this size for such a rare tumor can hardly be expected. In the discussions during the workshops it became evident, however, that this reflects above all the assessment of only two of the experts, while the other expert as well as the advisory expert panelist view the methodological weaknesses of this study to be too serious to place so much weight in the findings.

A study by Stang *et al.* [9] on uveal melanoma shows an increased odds ratio – the pooled estimate being 4.2 (95% CI 1.2–14.5) – for test subjects that are classified as likely exposed for at least half a year to mobile phones. The Auvinen *et al.* [10] study on brain tumors shows several methodological peculiarities (no selection bias, no recall bias) and two experts identify individual results as "noticeable". For instance, an odds ratio for developing brain tumors of 2.4 (95% CI 1.2–5.1) has been observed among users of analog mobile phones with contract durations of 1–2 years. The weak points of the study are the missing induction and latency time periods, and the poor estimate of exposure.

The study by Muscat *et al.* [11] demonstrates no general relationship between the usage of analog mobile phones and the risk of brain tumor development. This statement is limited, however, to an average cumulative exposure duration of only 2–3 years. Furthermore, the study did not consider latency periods for cancer development. In the assessment of the experts, a fractional analysis of the data on acoustic neuroma [12] is of only limited informational value because of the small number of cases and, therefore, very imprecise effect estimation.

The study of brain tumors by Inskip *et al.* [13] provides no indication of a hazard associated with the use of analog mobile phones for a few years. However, persons with longer exposure are only barely included, the latency period for cancer development is not considered, and the study focuses on too many and too small subgroups. In addition, multiple additional (however, only potential) risk factors are considered during model development through which the results are possibly biased. For these reasons, two experts assess the study as of limited reliability and therefore negative only to a limited extent. On the contrary, the third expert considers it as one of the qualitatively better and more reliable studies showing negative outcome regarding the brain tumor risk.

The Danish data from the international Interphone study shows no overall increased risk for development of acoustic neuroma [14]. [Interphone is an international, multi-center case-control study on brain tumor risks that is coordinated by the World Health Organization. It is conducted according to a jointly developed study protocol in identical manner in 13 countries with more than 7000 patients. Apart from Germany, the Scandinavian countries, Great Britain, France, Italy, Israel, Australia, New Zealand, Japan, and Canada are participants. Further information can be found in the Internet under: http://www.iarc.fr/ENG/Units/RCAd.html.] It should be noted, however, that an odds ratio of 0.26 was found for users whose first service contract was for an analog mobile phone. These persons have presumably used mobile phones for the longest period of time. Also, the value of 1.57 for the upper bound of the confidence interval of the ever/never analysis for mobile telephony contradicts a conclusion that mobile phone usage is associated with a significantly increased risk for acoustic neuroma.

The con argument is further supported by the retrospective cancer incidence study of Johansen *et al.* [15]. However, the conclusion for a strongly decreased relative risk for lung cancer (male relative risk = 0.65; 95% CI 0.58–0.73) cannot be judged as reliable since both selection bias and confounders were present. For brain tumors, selection bias and confounders have no significance; here, no increased risk is shown. This, however, is valid only for exposures to digital phones with usage of a few years (92% under 3 years).

According to the experts, an overall evaluation is quite difficult since most of the studies have methodological deficits and the room for interpretation of epidemiological studies is inherently wide.

The partially different interpretations of the study results among experts lead to different overall assessments. Considering the qualitatively good studies of Muscat, Inskip and the Danish part of the Interphone study, one expert believes that a carcinogenic effect of mobile telephony use is rather unlikely. However, this evaluation must be restricted to the relatively few tumor types that have been examined until now.

The overall evaluation of the other two experts is that a "vague initial suspicion" does exist for cancer risk from EMFs of mobile telephony. The experts view their conclusion supported by results of the studies from De Roos *et al.*, Auvinen *et al.*, Stang *et al.* and others. However, most of the positive test results refer to analog mobile phone technology, which is hardly used anymore. Based on these data, one expert cautiously concludes that, if – at all – a cancer-promoting effect from mobile telephony is present, it refers to analog mobile phone technology and not to the digital mobile phone system.

A consensus between the three experts was reached regarding the lack of increased risk for brain tumors among persons exposed to digital mobile phones for short durations (i.e. a few years). The experts have repeatedly presented very diverse uncertainties that widen the room for interpretation of epidemiological studies as well as those that constrain their validity considerably. The controversy of "significance testing versus estimation accuracy" (*cf.* [16]) belongs to the first category, while the second category includes the small case study numbers for subgroup comparisons as well as the assumptions of induction and latency periods. In particular, one expert also highlighted the problematic situation with exploratory studies which should not be used for the validation of hypotheses.

13.3
Insights from the Process of Building Evidence Maps

Evidence maps depict the result of a complex and time-consuming evaluation process, which usually involves various experts and comprises a number of different steps and workshop discussions. From accomplishing such a process within our research project, a number of insights can be drawn.

First of all, evidence characterization requires a valid input. A key issue is a thorough literature review. Usually this is not easily accomplished because research findings from numerous, very specific thematic fields must be evaluated. This requires knowledge and abilities that are best acquired through one's own research. Therefore, the process of evidence characterization should be carried out by experts from each of the research fields that are relevant for the evaluation. It certainly makes more sense to involve an epidemiologist when epidemiological studies have to be analyzed instead of relying on experts who claim to be generalists. In selecting the experts for the process of evidence characterization, it is important that the spectrum of different scientific positions is represented in the process of evidence characterization.

Another important point is the careful and plausible selection of the scientific literature, because only this will ensure that the experts will draw upon the same database. Hereby, the standards established for including or rejecting of scientific findings have to be outlined clearly in the expert opinion report. In our view, it is essential that independent advisory expert panelists participate, which themselves are not part of the scientific dispute over alleged health effects, but that have received recognition as experts for the respective research field. The inclusion of such consulting specialists is an important prerequisite for a competent, free of prejudice evidence evaluation.

Key to the process of constructing an evidence map is the systematic exploration of pro and con arguments for risks as well as the in-depth query of the underlying backings, i.e. the supporting and attenuating arguments. Therefore, it is crucial to make clear the standards, according to which contradictory findings are summarized into an overall picture, as well as to weigh fairly the pro and con arguments for the assessment. This requires, on the one hand, prudence and sensitivity and, on the other hand, the courage to challenge the experts because it is essential for the preparation of the final risk characterization that the assessments are not biased and plausibly justified.

13.4
Conclusions

A number of approaches to characterize and grade levels of evidence are being used in risk assessment. The evidence map approach aims primarily at a clear and transparent characterization of evidence for causality so that the scientific status of a potential hazard can be better understood. This, in turns, can help people to make informed decisions. However, at least two problems require further elaboration and research.

First, evidence maps in their present form do not provide a categorization or grading of evidence. In fact, the reason for developing evidence maps was to provide a "richer" context of characterizing evidence as compared to the category labels of, for instance, the EPA [17] or International Agency for Research on Cancer ([18], see also Chapter 9 of this volume). Apart from its structure, which is the same for all evidence maps, the content of an evidence map is quite close to the underlying expert assessments, often referring to brief citations or paraphrasing statements from the experts' reports. So far, no standardized terminology exists which makes it difficult to compare the state of evidence across different evidence maps. Specifically for the "conclusions" section, it would be meaningful to develop a grading system using standardized terms to describe the conclusions. The format used in the California EMF Program (see Chapter 12 of this volume) is certainly an interesting candidate; however, other formats might be worth considering (e.g. [19]).

Second, it remains an empirical question whether such an evidence map will attain the intended objectives, and whether or not evidence maps will perform better than other approaches for evidence characterization. Therefore, further research is needed to evaluate the communicative strengths and weaknesses of evidence maps.

References

1 Toulmin, S. (1958) *The Uses of Argument*, Cambridge University Press Cambridge.

2 Toulmin, S., Rieke, R. and Janik, A. (1984) *An Introduction to Reasoning*, Macmillan New York.

3 Krimsky, S. (2005) The weight of scientific evidence in policy and law, *American Journal of Public Health*, **95** (Supplement 1), S129–S136.

4 Linkov, I. and Satterstrom, F.K. (2006) Weight of evidence: what is the state of the science? *Risk Analysis*, **26**, 573–575.

5 Weed, D.L. (2005) Weight of evidence: a review of concept and methods, *Risk Analysis*, **25**, 1545–1557.

6 EPA (2004) *An Examination of EPA Risk Assessment Principles and Practices (EPA/100/B-04/001)* US Environmental Protection Agency, Washington, DC.

7 Blettner, M., Jöckel, K.-H. and Stang, A. (2005) Epidemiologie Krebs. in *Risikobewertung Mobilfunk: Ergebnisse eines wissenschaftlichen Dialogs*, (eds P.M. Wiedemann, H. Schütz and A. Spangenberg), pp. C1–C44, Schriften des Forschungszentrums Jülich, Jülich.

8 De Roos, A.J., Teschke, K., Savitz, D.A., Poole, C., Grufferman, S., Pollock, B.H. and Olshan, A.F. (2001) Parental occupational exposures to electromagnetic fields and radiation and the incidence of neuroblastoma in offspring, *Epidemiology*, **12**, 508–517.

9 Stang, A., Anastassiou, G., Ahrens, W., Bromen, K., Bornfeld, N. and Jockel, K.H. (2001) The possible role of radiofrequency radiation in the development of uveal melanoma, *Epidemiology*, **12**, 7–12.

10 Auvinen, A., Hietanen, M., Luukkonen, R. and Koskela, R.S. (2002) Brain tumors and salivary gland cancers among cellular telephone users, *Epidemiology*, **13**, 356–359.

11 Muscat, J.E., Malkin, M.G., Thompson, S., Shore, R.E., Stellman, S.D., McRee, D., Neugut, A.I. and Wynder, E.L. (2000) Handheld cellular telephone use and risk of brain cancer, *Journal of the American Medical Association*, **284**, 3001–3007.

12 Muscat, J.E., Malkin, M.G., Shore, R.E., Thompson, S., Neugut, A.I., Stellman, S.D. and Bruce, J. (2002) Handheld cellular telephones and risk of acoustic neuroma, *Neurology*, **58**, 1304–1306.

13 Inskip, P.D., Tarone, R.E., Hatch, E.E., Wilcosky, T.C., Shapiro, W.R., Selker, R.G., Fine, H.A., Black, P.M., Loeffler, J.S. and Linet, M.S. (2001) Cellular-telephone use and brain tumors, *New England Journal of Medicine*, **344**, 79–86.

14 Christensen, H.C., Schuz, J., Kosteljanetz, M., Poulsen, H.S., Thomsen, J. and Johansen, C. (2004) Cellular telephone use and risk of acoustic neuroma, *American Journal of Epidemiology*, **159**, 277–283.

15 Johansen, C., Boice, J., Jr., McLaughlin, J. and Olsen, J. (2001) Cellular telephones and cancer – a nationwide cohort study in Denmark, *Journal of the National Cancer Institute*, **93**, 203–207.

16 Rothman, K.J. and Greenland, S. (1998) Approaches to statistical analysis. in *Modern Epidemiology*, 2 edn, (eds K.J. Rothman and S. Greenland), pp. 183–199, Lippincott Wilkins & Wilkins, Philadelphia, PA.

17 EPA (2005) *Guidelines for Carcinogen Risk Assessment*, review draft ed., US Environmental Protection Agency, Washington, DC [http://www.epa.gov/ncea/] [Retrieved: 24.07.2007].

18 IARC (2006) *Preamble to the IARC Monographs. IARC Monographs Programme on the Evaluation of Carcinogenic Risks to Humans*, International Agency for Research on Cancer, Lyon [http://monographs.iarc.fr/ENG/Preamble/CurrentPreamble.pdf] [Retrieved: 24.09.2007].

19 Schünemann, H.J., Best, D., Vist, G. and Oxman, A.D. (2003) Letters, numbers, symbols and words: how to communicate grades of evidence and recommendations, *Canadian Medical Association Journal*, **169**, 677–680.

IV
Psychological and Ethical Aspects in Dealing with Conflicting Data and Uncertainty

14
Perception of Uncertainty and Communication about unclear Risks
Peter Wiedemann, Holger Schütz, and Andrea Thalmann

14.1
Introduction

Scientific uncertainty represents a major challenge for risk communication [1, 2], and for more than 20 years researchers from different fields have continued to address the question of how to effectively communicate uncertainties and how to deal with and enhance lay peoples' understanding of uncertainty, e.g. in the processing of probabilistic information [3].

In this chapter we will review the literature on communication and perception of uncertainty from the particular perspective of how communication can help to improve the understanding of unclear risks. We characterize unclear risks herein as risks that are not proven, but cannot be excluded, i.e. risks that are by definition uncertain.

To help nonexperts understand this particular type of risk, at least three points must be addressed. First, it is necessary to provide for a basic understanding of the overall impact of uncertainty on the risk assessment process. Second, nonexperts should become aware that risk assessments can in fact never be perfect because they inevitably entail knowledge gaps due to such challenges as measurement problems, lack of data or limited understanding of the mode of action by which a particular agent might induce an adverse health effect. Third, nonexperts should receive information about which sources of uncertainty are critical as well as information about the remaining evidence on which the risk characterization is based.

We structured the present chapter into four main parts. Following this introduction, we provide a conceptual background and discuss uncertainty in the context of risk assessment. The next part presents an overview of the existing literature on perception and communication of uncertainty information and related issues. It is followed by a section on the perception of precautionary measures. Finally, we discuss the open questions for communicating unclear risks and offer an outlook of further research needs. In addition, practical implications for risk communication are outlined.

The Role of Evidence in Risk Characterization: Making Sense of Conflicting Data.
Edited by Peter M. Wiedemann and Holger Schütz
Copyright © 2008 WILEY-VCH Verlag GmbH & Co. KGaA, Weinheim
ISBN: 978-3-527-32048-6

14.2
Uncertainty in Risk Assessment

Risk assessment of substances or agents that are suspected to pose a public health problem is a science-based process typically comprising four steps: hazard identification, dose–response assessment, exposure assessment and risk characterization [4]. The aim is a statement regarding the probability of exposed persons being harmed. However, the quantitative results of risk assessments are often inconclusive. Risk assessments may suffer from measurement errors, data gaps, lack of theoretical understanding of the cause–effect relationship, as well from extrapolation problems. Uncertainty is therefore an inevitable component of any risk assessment.

There is a vast body of literature that has tried to classify different types of uncertainties (e.g. the taxonomies of uncertainty by Finkel [5], Morgan and Henrion [6], and the National Research Council [7]). Even a superficial glance reveals that the extent of uncertainty may vary widely in degrees of certainty, starting from being almost certain up until positing mere speculations. However, uncertainty may also refer to different aspects of the risk assessment process. It can equally be found to be a part of the hazard identification, dose–response assessment or exposure assessment.

The US Environmental Protection Agency handbook on risk characterization stresses the point ([8], p. 42):

> Because every risk assessment has many uncertainties, and involves many assumptions, the challenge in characterizing risk for decision makers, whose time is limited and who are not risk experts, is to convey that small subset of key findings and strengths and limitations that really makes a difference in the assessment outcome.

One of these key issues that must be addressed in order to help nonexperts understand the findings of a risk assessment refers to a basic distinction between hazard and risk. Accordingly, there are two different types of uncertainties. The first type is uncertainty with respect to the hazard ("Is the hazard real?") and the second type is uncertainty with respect to the magnitude of the risk [9]. [According to the International Programme on Chemical Safety hazard is defined as the "Inherent property of an agent or situation having the potential to cause adverse effects when an organism, system, or (sub)population is exposed to that agent" ([9], p. 12). In contrast, risk refers to the "probability of an adverse effect in an organism, system, or (sub)population caused under specified circumstances by exposure to an agent" ([9], p. 13).] The former type of uncertainty deserves special attention because it concerns the heart of the risk assessment process regarding unclear risks. Namely, what is known about the capability of a suspected hazard for causing adverse health effects in humans; in other words, does the available scientific evidence support the hazard assumption. The implication is self-evident that the hazard has to be clearly identified before any reliable statement about risks can be made. To bring it to a point, if there is no hazard, there is no risk at all. Hazard identification requires that the relevant studies consistently indicate an established effect, i.e. a coherent relationship between exposure and the endpoint. Furthermore, the endpoint should refer to an unequivocally detrimental effect.

The matter becomes even more complicated if one takes into account the various types of research that have to be integrated into the overall scientific picture. Typically, epidemiological studies on humans, laboratory animal studies, as well as studies on cells and tissues have to be assessed. It is not unusual that these different study types yield different results.

For any and all of these cases, the existing uncertainty may stem from a lack of information or it may be the result of disagreement among experts about what is known and what is unknown, and how to interpret the available knowledge. Nevertheless, key issues are that quality and quantity of uncertainty in hazard identification may vary and therefore hazards might be categorized according to their level of uncertainty [10]. These issues challenge the communication skills of the risk assessors for delivering a report about the hazard identification that is fair and transparent as well as reasonable and understandable.

In their report addressed to a nonexpert audience, risk assessors have to make clear whether they have identified a hazard beyond a shadow of a doubt. If they are uncertain about the existence of a hazard, risk assessors should provide information on the underlying evidence that speaks for a hazard and delineate the level of the remaining uncertainty.

For instance, Moss and Schneider ([11], p. 37) have proposed seven recommendations for the third assessment process of the Intergovernmental Panel on Climate Change to handle uncertainties in order to provide more consistent assessment and reporting. From our perspective, the most significant proposals are:

- Identify the most important factors and uncertainties that are likely to affect the conclusions
- Document key causes of uncertainty
- Determine the appropriate level of precision of your conclusions after considering the nature of the uncertainties and state of science
- Use standardized terms for describing the state of scientific information and level of uncertainty on which the conclusions and/or estimates are based
- Prepare a "traceable account" of how the estimates were constructed

Moss and Schneider ([11], p. 35) argue that:

> ... it is more rational for scientists debating the specifics of a topic in which they are acknowledged experts to provide their best estimates of probability distributions and possible outliers based on their assessment of the literature than to have users less expert in such topics make their own determinations.

In other words, experts should help lay people make informed judgments.

From a psychological perspective these recommendations on how to communicate uncertainties are prescriptive advice that appears to fit the theoretical understanding of the problem. The key issue, however, is whether they are grounded in empirical research. More specifically, the open question remains whether this advice will help nonexperts reach the right conclusions from uncertainty information in order to make informed judgments. This issue will be analyzed in the following section.

14.3
Uncertainty Communication and Lay Persons' Perception of Uncertainty Information

In a survey on information desire about new risks that are unknown to lay people, Lion *et al.* [12] demonstrated that these persons are most interested in three questions: "How is one exposed to the risk?", "What does the risk mean?" and "What are the consequences?". Information about the probability of the risk seems to play a less important role. This seems to indicate – in the view of the study authors – that people want to know first what the risk actually results in (i.e. its manifestation), in order to assess its relevance. They are first of all interested in information explaining the nature of the risk. Consequently, it can be assumed that people are also quite interested in information about uncertainty aspects of the risk assessment.

However, the communication of uncertainty in risk assessment is confronted by several challenges. In general, lay people have difficulties in making sense out of risk assessment data. Thus, the question arises, how one can enhance lay people's background knowledge necessary to interpret risk assessments. Furthermore, lay people are typically not acquainted with the fact that uncertainties inherently influence risk assessments. Consequently, the challenge for risk communicators is to explain the underlying uncertainties in any risk estimate. Finally, lay people do not have an appreciation of how uncertainty influences the quality of a risk assessment. Therefore, ways have to be found to raise lay peoples' ability for critical appraisal.

The following review of the empirical research on uncertainty information and communication is divided into five parts. The first part highlights how lay people comprehend the risk assessment framework. The second part provides an overview of the research on how lay people understand information about relative risks. The third part presents research on how informing about uncertainties in risk assessment will impact risk perceptions, and the fourth part outlines the research on lay people's understanding of verbal, numerical and graphical expressions of uncertainty in risk assessment. Finally, the fifth part focuses on the role of context in communication of uncertainties.

14.3.1
Intuitive Toxicology: How do Nonexperts Understand the Risk Assessment Framework?

The risk perception literature clearly indicates that lay people and experts differ with respect to their risk appraisals. The difference is caused by a variety of reasons; one of them refers to the limited familiarity of the lay public with formal, scientific risk assessment procedures. This issue was not addressed until Paul Slovic and his colleagues conducted several studies on intuitive toxicology [13–17]. These surveys took place in Canada, the UK and the US. The fundamental goal was to clarify how

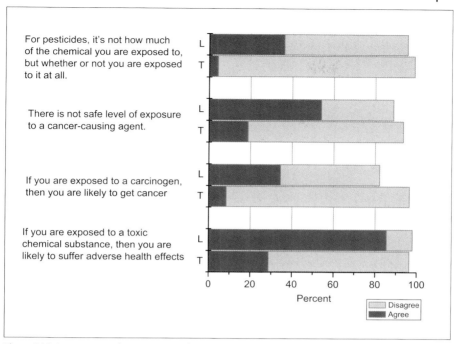

Figure 14.1 Intuitive toxicology: responses from lay people (L) and from toxicologists (T) (13, p. 217).

nonexperts interpret basic toxicological concepts and assumptions as well to assess opinions about and attitudes towards chemical risks. The studies focused on hazard evaluation, dose–response sensitivity, exposure issues, and trust in animal and "bacterial studies" (this is the original expression used in the papers on intuitive toxicology – it seems that the authors mean *in vitro* studies, such as the Ames test [18]).

Overall, the data suggest that nonexperts and experts differ in a variety of issues (Figure 14.1) and also point to several misperceptions by nonexperts. To name only the most important: lay people are less sensitive than experts in their appraisal of risks to the amount of exposure and to dose–effect relations.

The study of MacGregor *et al.* [17] also provides insights into the intuitive concept for cancer-causing circumstances, i.e. what do people have in mind when they are informed that a particular chemical can cause cancer? When challenged with descriptive statements about a particular chemical, but without the necessary information to reach an informed decision, over 80% of respondents intuitively agreed with the statement "The chemical has properties which make it possible for it to cause cancer". However, most people do take into account the amount of exposure, the circumstances of exposure as well as differences in the individual susceptibility. More than 90% of the respondents disagree with the statement "The chemical always causes cancer in everyone exposed, no matter how small the exposure". On the other hand, it seems that people draw strong inferences when they are informed that they are exposed to a chemical that can cause cancer: 70% infer that exposure is serious, 50% believe that many people are exposed and 58% conclude that the exposure probably results in cancer. (the original text of the statement was: "People in Oregon are exposed to chemicals that can cause cancer"; note that the study was conducted with students from the University of Oregon)

Further data indicate that lay people are divided in their answers to the question "If a person is exposed to a chemical that can cause cancer in humans, then the person will probably get cancer some day". About 43% agreed with this statement and about 44% disagreed with it. About 13% did not know. When asked about their opinion on the statement "If a person is exposed to an extremely small amount of a chemical that can cause cancer, then the person will probably get cancer some day", the majority of the respondents (about 80%) disagreed [17]. Furthermore, the same study reveals that people have different views with respect to whether there is an exposure level below which a cancer causing chemical does not cause cancer (about 50% agree, about 28% disagree and 22% do not know). According to the study authors, many respondents were either inconsistent or unable to express an opinion. They conclude that even subtle changes in how the concept of exposure is communicated might evoke quite different beliefs [17].

A further interesting aspect is the appraisal of animal studies in cancer research. It seems that most lay people consider animals' biological responses to a chemical as a valid indicator of how the human body would react to the same chemical. About 79% of the respondents agree with the statement "If a scientific study produces scientific evidence that a chemical causes cancer in animals, then we can be reasonable sure that the chemical will cause cancer in humans" ([13], p. 218). However, the respondents are divided in their opinions on the significance of "bacterial studies" for cancer research.

The studies on intuitive toxicology indicate that public controversies about chemical risks are amplified by disagreements among experts and the limitations of risk assessments. These limitations are also addressed by Purchase and Slovic [19]. In a risk perception study the authors compared two methods used in risk assessment; the safety factor approach based on the No Observed Adverse Effect Level (NOAEL) and on the nonthreshold model. They demonstrated that risk assessments based on a nonthreshold model of carcinogenesis amplifies public fears.

14.3.2
How do People Understand Information about Relative Risks

The relative risk is an indicator often used in epidemiology as well as in animal research in order to characterize the magnitude of a risk factor. Usually, relative risk is defined as the probability of occurrence of an adverse effect in subjects who are exposed to a risk factor compared to subjects without such an exposure. The relative risk therefore presents the relative size of an effect. It indicates the increased or decreased frequency of a certain adverse effect in exposed subjects compared to unexposed subjects. For instance, a relative risk of 1.5 means that in the exposed group the risk of the adverse effect is 50% higher than in the nonexposed group.

Psychological research is interested how nonexperts interpret such relative measures compared to other statistical information formats. Gyrd-Hansen *et al.* [20] asked their Danish study participants to state their preferences for risk-reducing health care interventions based on information on absolute risk reduction (ARR) or relative risk reduction (RRR). A preference for increases in RRR was demonstrated. There was a stronger tendency to choose the intervention that offered the highest RRR, if the RRR was explicitly stated.

Studies suggest that relative risk information has a stronger impact on risk perceptions than base rate information [21]. The relative risk description leads to an increase in risk perception compared to a risk description based on incidence/base rate data. To give an example, a 100% higher risk of dying could be equivalent to an increase from one to two fatalities out of 1 000 000 people. However, the first description sounds much more alarming then the latter. Therefore, it seems to be advisable to take into account the incidence rate or base rate when interpreting a relative risk. Nevertheless, base rate information is often neglected ([22]; however, see Ref. [23] for a skeptical view as to whether the base rate neglect as a general phenomenon actually exists). Lay persons tend to ignore the base rate intuitively and use more case-specific information while interpreting the probability of an event. For example, people watching the TV weather forecast typically focus so much on the rainfall probability information that they forget the general frequency of raining, i.e. the base rate. However, this seems to be an error ([24], p. 776):

> With forecast accuracies of 83%, one might expect that a forecast of rain during the walk would prove correct 83% of the time. However, the hourly base-rate of rain in the UK is so low that forecasts of rain are over twice as likely to be wrong as right ...

14.3.3
Information about Uncertainty in Risk Assessment: How do Nonexperts Cope With It?

The research on uncertainty has a long and celebrated history. Following the seminal work of Ellsberg [25] on ambiguity (i.e. uncertainty about probability) many economists and psychologists have analyzed the so-called Ellsberg Paradox, i.e. the aversion of ambiguity in lotteries and other choice tasks [26]. It seems that people prefer known risks, i.e. risks with known probabilities, over unknown risks, i.e. risks for which the probabilities are unknown.

The first paper addressing how lay people assess uncertainties in chemical risk assessment was written by Branden Johnson and Paul Slovic in 1995 [27]. It explores how people interpret and evaluate information about uncertainties in risk assessment. Their results show that being informed about uncertainties leads to different appraisals. For some study participants, the disclosure of uncertainty results in an increase of trust in the information source, while others consider this as a sign of incompetence or dishonesty. In the same study, the presentation of confidence intervals for risk estimates leads to a higher perception of risk than when point estimates were used. A further study of the perception of uncertainty in risk assessments demonstrated that lay people attribute uncertainty to social factors (e.g. self-interest or incompetence of the experts) and not to features of the risk assessment (e.g. complexity of the task or limited resources) [28]. Furthermore, the study authors found that the upper limit of confidence intervals is seen to be the most credible estimation. About half of the study participants prefer an "either/or" communication, i.e. a clear message of the existence or nonexistence of a risk. The study of Johnson [29] largely replicates these results. The findings show that general attitudes and evaluations towards the information provider had an influence on the appraisal of the uncertainty. Those who were more mistrustful were more inclined to consider the statement of uncertainty as a sign of dishonesty and incompetence.

Kuhn [30] has analyzed the effects of uncertainty information on perception of environmental risks. For five different hypothetical scenarios she presented risk information in four message formats: (i) a single estimate, (ii) a single estimate with a verbal description which attributed the uncertainty to incomplete knowledge, (iii) a numerical range centered around the estimate without any further information about the nature of the uncertainty or (iv) a numerical range plus background information which explains that the different values are the result of two risk assessments from organizations with different values and biases. After reading the scenarios the study participants were asked to rate the riskiness of the hazards and their general personal environmental concern. The results suggest that the message format has no significant effect on risk perception. However, people with higher concerns perceived greater risks. The data indicate also an interaction effect. It seems that for concerned people the message format has a significant impact on risk perception, in contrast to the unconcerned people. Higher concerned people show significantly higher risk perception in the point estimates condition and in the numerical range plus background information condition.

Miles and Frewer [31] have conducted a study on public perception of scientific uncertainty in relation to five different food hazards. They studied the effects of seven types of uncertainties on risk perception: uncertainty about whether a hazard exists (in terms of the authors: "temporal uncertainty"), uncertainty about who is affected, measurement uncertainty, uncertainty due to disagreement of experts, extrapolation uncertainty, uncertainty about the magnitude of a risk and risk management uncertainty. Study participants had to evaluate the uncertainty statements, e.g. "Whilst scientists and risk regulators (e.g. the government) believe that there is a health risk from [bovine spongiform encephalopathy], they are uncertain who in the population is affected". Each statement had to be rated with respect to seriousness of the risk in both personal and societal perspective.

The findings suggest that nonexperts focus primarily on the type of hazard. Their ratings of the seriousness of the risk mainly depend on the source, e.g. the *Salmonella* hazard reached higher ratings for all uncertainty statements than the pesticides hazard. With respect to the type of uncertainty it seems that the study participants are more concerned about uncertainty in risk management, measurement uncertainty and uncertainties about the size of the risk. Perhaps it is not that surprising, but certainly disappointing, to find that uncertainty with respect to hazard identification does not play an important role. However, a factor analysis resulted in only one factor, i.e. people do not clearly differentiate between the various types of uncertainties. They respond to the different uncertainties in a uniform way.

Ogden *et al.* [32] analyzed how doctors' expression of uncertainty might affect their patients. The patients were asked how confident they would feel if their doctors used various uncertainty expressions, like "I'm not sure about this". Additionally, doctors were asked how confident they thought the patient would be if they used the expressions. Their study revealed considerable differences between the GPs' and patients' judgments. Expression of uncertainty can unnerve patients, and doctors tend to underestimate this negative effect.

The available studies appear to support the assumption that nonexperts have difficulties to differentiate among different types of uncertainty and to draw the appropriate conclusions for their risk appraisal.

14.3.4
Uncertainty Descriptions: How do People Understand Qualitative, Quantitative and Visual Expression?

The literature on uncertainty description reveals different approaches that are used to describe uncertainties in risk assessments. Uncertainty can be described qualitatively by verbal phrases (so called hedging phrases) or by a standardized set of verbal labels, or quantitatively, i.e. confidence intervals or probability distributions. Furthermore, uncertainty can be expressed by graphical means. In the following discussion we will focus on these approaches. Please note, however, that the advanced statistical methods will be not considered because they do not play a role in the characterization of unclear risks.

Verbal hedging phrases (e.g. it seems, tentatively, possibly, probably, etc.) are used when scientists intend to avoid too strong statements that are not covered by the data [33]. In risk assessments, hedging phrases are employed if a hazard has neither been unequivocally proven nor can be unequivocally rejected. (Such expressions are also used when a speaker interested in a communication free of conflict wishes to avoid making himself/herself vulnerable to attack. They are verbal protective shields.) Risk assessors are forced to deploy such phrases in order to take into consideration knowledge gaps and contradictory results. Often the use of hedging phrases is a matter of subjective judgment to avoid overconfidence and capitalization of chance (to see signals even in random patterns). However, vast room for interpretation exists in the meaning of these phrases. What for instances the phrase "rather unlikely" means in contrast to the phrase "relationship can not be excluded" might be very differently interpreted.

There are only few studies on hedging phrases (e.g. [34]). Unfortunately, no systematic empirical study exists that investigates the impact of these phrases on risk perception. However, other research suggests that those subtle phrases might have a remarkable impact. A striking example is an experiment on the influence of reasons on the interpretation of probability forecasts [35]. The results suggest that positive reasons (e.g. you have large veins, it will make the surgery easier) given for a probability estimate regarding the failure of a surgery elicit a more optimistic view than negative reasons (e.g. you have small veins, it will make the surgery more difficult) given for the same probability estimate.

As mentioned above, uncertainty is sometimes expressed by set of standardized verbal labels. For instance, the International Agency for Research on Cancer (IARC) [36] differentiates among five level of evidence for carcinogenicity that are labeled as Group 1 ("The agent is carcinogenic to humans"), Group 2A ("The agent is probably carcinogenic to humans"), Group 2B ("The agent is possibly carcinogenic to humans"), Group 3 ("The agent is not classifiable as to its carcinogenicity to humans") and Group 4 ("The agent is probably not carcinogenic to humans"). The categorization into one of these five categories is based on specified rules ([37] and Chapter 9 of this volume). Compared to the use of idiosyncratic hedging phrases the IARC approach is a step forward because it narrows the range of interpretations by using a common language. Nevertheless, the issue remains open as to how lay people interpret such standardized phrases, e.g. if they correctly infer the difference between "possible" and "probable". The research question to address here refers to the lay people's understanding of qualitative likelihood expressions. Usually this kind of research is based on a translation task, i.e. study participants are asked to assess the numerical probability of various uncertainty terms [38].

Several empirical studies reveal the ambiguity of verbal quantifiers [3, 39–45]. The subjective meaning of verbal expression using probability or frequency indicators such as "probable", "frequent", "possible" or "often" differ from person to person. Furthermore, the lay persons' understanding of verbal uncertainty categories does not necessarily correspond with the experts' intended statement. As Fischer and Jungermann [44, 45] demonstrate in their experiment on the understanding of verbal labels that indicate the frequency of occurrence of pharmaceutical side effects, the lay persons' numerical equivalents of verbal labels (rarely, occasionally and frequently) differ form the numerical definition given by the German Health Agency for the three labels.

Similarly, Jablonowski [46] demonstrates how variable numerical interpretations of verbal quantifiers are across people. An example here is the meaning of the verbal probability label "unlikely": Asking people to equate a numerical value for "unlikely" on a scale from 0.00 to 1.00, the result will be a broad range from 0.09 to 0.30. Additionally, the interpretation of "unlikely" will be quite similar to the estimation for "somewhat likely" (which range from 0.09 to 0.45).

Furthermore, research has indicated that verbal expressions are less neutral than numerical statements. Therefore, they might influence judgments in subtle ways [3, 38, 47]. To name only one example, there are expressions that accentuate positively the existence of an occurrence, such as "possibly", while others like "doubtful" emphasize their nonexistence [48, 49].

In particular, the meaning of verbal uncertainty categories depends highly on the *context* in which they are embedded [39, 40, 43, 50–52]. Contextual characteristics modify the meaning of verbal quantifiers. For instance, the numerical equivalent of verbal expression of frequencies such as "rarely", "sometimes" or "frequent" depends on the specific context in which these words are used. The following example shows this dependence very nicely. The lay persons' numerical interpretation of "frequent" will be quite different in the two following cases: "Going to the movie theatre frequently" and "Visiting Asia frequently". It demonstrates how contextual characteristics might modify the meaning of verbal uncertainty descriptions. In fact, many experiments demonstrate that verbal probability descriptions of severe events are interpreted numerically higher compared to verbal probability description of less severe events (e.g. [42]). People are more likely to choose more certain sounding probability expressions with respect to events which have more serious consequences. In other words, people confound probability with the magnitude of harm.

A study by Thalmann [53] tested evidence categories with regard to their clarity and comprehensibility. The results of her study suggest that such verbal evidence descriptions are highly ambiguous and will be understood in different ways from person to person. Asking a lay person to associate different verbal evidence categories to numerical values on a scale form 0% power of evidence to 100% power of evidence brought up disconcerting findings, the lay persons' estimations appear widely spread in each verbal evidence description and overlap with other descriptions of the power of evidence, thereby contradicting themselves. For example, the numerical estimation of the verbal evidence expression "consistent hint" is spread between 6 and 100% power of evidence and is almost identical with the numerical estimation of the verbal evidence description "strong hint" (between 16 and 90%).

While idiosyncratic and standardized verbal uncertainty descriptions remain purely qualitative, other approaches have attempted to summarize and characterize uncertainty in a (semi-)quantitative manner. For instance, within the California Electromagnetic Field (EMF) Program ([54] and Chapter 12 of this book), program staff experts were asked to express their "degree of certainty" about a causal relationship between (low-frequency) EMF exposure and effects on selected endpoints. These judgments were based on a thorough review of the available experimental and epidemiological evidence and on an extensive elicitation procedure which included a discussion of how much more (or less) likely a specific pattern of evidence would be under the causal hypothesis compared to the noncausal hypothesis. The "degree of certainty" was expressed on a numerical scale ranging from 0 to 100; in addition the experts also assigned a confidence interval around their point estimates. Ranges of the scale were also linked to verbal phrases, e.g. "90 to 99.5: strongly believe that they increase the risk to some extent".

With respect to this type of semiquantitative uncertainty characterization another stream of research becomes important, i.e. how numerical information, usually expressed in probabilities or probability confidence intervals, is understood.

Here key issue is "innumeracy", i.e. the inability to make sense of the numbers. As research shows, lay people in particular have difficulties with probabilistic reasoning. They often fail to weight small probabilities with potential consequences correctly [55–58]. The Prospect Theory by Kahneman and Tversky [59] indicates that small probabilities are overweighted and high probabilities are underweighted.

As already mentioned above, the study of Johnson and Slovic [28] suggests that lay people appraise the upper limit of confidence intervals as the most credible estimation. Viscusi [60] came to the same conclusion. Schapira *et al.* [61] conducted a study with female patients. Less-educated women perceived confidence intervals as "wishy-washy", whereas better-educated women preferred this uncertainty information. The finding suggests that the effectiveness of a format not only depends on the characteristics of the information, but also on cognitive characteristics of the recipients.

Visual aids seem to be a promising way to facilitate the interpretation of complex information about uncertainties. Two streams of research appear to support this assumption. First, during recent decades several researchers have studied how graphical displays help patients to understand probabilistic information. Second, there are also a select number of studies available focusing on how to communicate uncertainty information with respect to risk assessments. However, a detailed discussion of graphic risk/uncertainty communication is beyond the scope of this paper.

In the clinical context, researchers have studied how to communicate risk information to patients (for an overview, see Lipkus and Hollands [62]). For instance, clinicians use various graphical representations of probabilities for risk communication, e.g. bar charts, thermometer graphs, icons, lines curves and survival curves. There is evidence that vertical bars together with numerical estimates may improve the understanding of probabilities [63]. Illustrations such as cartoons appear to aid understanding of probability information as well. However, it should not be taken for granted that visual aids are always more intuitive than verbal or quantitative expressions; many studies found that patients' interpretations of graphical risk displays were dependent upon expertise or instruction [64].

Research has paid only little attention to the use of graphics in communicating uncertainties of risk assessment information. Since the early study of Ibrekk and Morgan [65] only a few papers have addressed this issue. One of these is the study by Thompson and Bloom [66]. In an investigation based on focus groups with risk managers, their findings suggest that risk managers prefer simple charts and graphs without too many details. Furthermore, their study participants preferred a probability density function in order to visualize uncertainties. However, Ibrekk and Morgan [65] argue that a cumulative distribution function has some advantages in displaying uncertainties. Therefore, they recommend presenting both functions together.

14.3.5
Contextual Effects

It is scientific common sense that understanding as a result of verbal interaction is not only a mere translation of spoken words into the seemingly connected meanings [77]. On the contrary, understanding is an active process using many cues including own prior beliefs.

Fox and Irwin [68] provide a very helpful overview on the role of context in the communication of uncertain beliefs. Their model highlights three main aspects of context. The first aspect refers to the listeners' prior beliefs as well to their interpretation of the discourse context. The second covers the listeners' view of the speaker's intention and the evaluation of the speaker's credibility. The third feature relates to the listener's understanding of the speaker's statement.

Understanding of uncertainty information seems to be more than just a simple process of decoding numbers or verbal quantifiers. First of all, prior beliefs of the listeners play a crucial role because humans evaluate new information about uncertainties in the light of their own belief system. Fox and Irwin [68] argue that the understanding of qualitative probability quantifiers (e.g. rare or possible) describing the likelihood of a certain event depends also on prior beliefs on the probability of this event. In their view, the interpretation of probability quantifiers follows the Bayesian model where prior beliefs are updated in the light of new data. They further suggest that the willingness to update beliefs about probabilities is a critical variable because it seems rational to change one's own position in the light of new information. However, two biases might occur. First, researchers have observed a negativity bias, i.e. the preference for negative information [69], and it has been indicated that negative information elicit more rapid and more prominent responses than nonnegative information [70]. Furthermore, is has been shown that the context, e.g. prior beliefs and attitudes, play a significant role in evaluating new information. For instance, Koehler [71] has found – both in a laboratory experiment with graduate students and a quasi-experimental survey with practicing scientists – that scientific judgments of the quality of research reports were influenced by whether the reports confirmed or disconfirmed the subjects' prior beliefs. When research reports confirmed the prior beliefs they were judged to be higher quality than those who did not agree with these beliefs. Studies also have typically found that people are more likely to seek out and attend to data which are consistent rather than data which are inconsistent with their initial beliefs [72].

Wiedemann et al. [73] studied the influence of prior beliefs in an EMF risk perception survey. They constructed different scenarios outlining possible critical developments or changes in the debate on risks from cellular phones and base stations. Half of the scenarios describe warnings, i.e. risk confirming news. The other half of the scenarios consists of reassuring messages, i.e. risk disconfirming news. The study participants were asked to indicate their willingness to change their own risk perception in the light of each of these several scenarios on a seven-point rating scale.

The study results indicate an asymmetry between warnings and reassuring scenarios with respect to the impact on one's willingness to alter risk perceptions. In average, warning scenarios turned out to be much more influential than of reassuring scenarios. Most interestingly, this effect is stronger for the concerned compared to the unsure and unconcerned people.

Smithson [74] suggests that the perceived source of uncertainty is an additional key variable. It matters whether the uncertainty is caused by conflicting evidence or whether the uncertainty is consensually perceived. His results suggest that the conflicting situation causes higher perceived uncertainty than the consensus situation. Further research has shown that risk perceptions based on experiences are in many respects different form risk perceptions based on abstract statistical information [75]. The bottom line seems to be that risk perception, if grounded in experience, is much more persuasive and more difficult to change. It is plausible to assume that uncertainty information that collides with experience-based beliefs will be mostly neglected.

Moreover, the interpretation of uncertainty information depends on how the messenger is perceived. Here, trust is the key variable. According to the Trust, Confidence and Cooperation model, developed by Earle *et al.* [76], trust is the willingness – in expectation of beneficial outcomes – to make oneself vulnerable to another, based on a judgment of similarity of intentions or values. If people have no or rather low trust in an information source then they doubt the validity of the message delivered by the information source [77]. In agreement with this observation, White *et al.* [78] have shown that people have greater trust in messages that are consistent with their prior beliefs.

Furthermore, the appraisal of uncertainty information may be influenced by the various inferences regarding the information sources, e.g. its alleged motivation, its perceived degree of accountability, as well as its supposed goals [68]. Other context factors might also play a significant role. Zukier and Pepitone [79] have demonstrated that slight changes in the interpretation of the social context will influence the processing of probability information. Their study participants were more likely to use base rate information when instructed to behave as statisticians than when instructed to imitate clinical psychologists. Similar insights have been found by Schwarz *et al.* [80] and Dennis and Babrow [81].

14.4
Lay Peoples' Perception of Precautionary Measures

A crucial issue in many debates on the management of potential environmental health risks is the controversy about precautionary measures. This controversy centers around whether, when and which precautionary measures should be invoked in order to mitigate possible, but as yet unproven hazards.

While many international regulatory governance bodies, such as the World Health Organization and the World Trade Organization, are still formulating their guidelines for a rational application of the precautionary principle in the environmental health field, a number of European countries, ostensibly empowered by the European Community precautionary principle communication of 2000, have already implemented precautionary policies not only to prevent the potential for harm, but also as a management approach to cope with public concerns of high-profile risk issues.

With respect to risk perception, one crucial question arises: "Do concerned people feel safer when they know that precautionary measures are in place to protect their health?". Basically, two outcomes can be envisioned. On the one hand, informing the public about the implementation of precautionary measures may strengthen trust in the regulatory bodies and the industry, and thus reduce risk perception and possibly even mitigate public outrage. On the other hand, disclosure of precautionary measures may result in countervailing risk perception effects, e.g. undermine the credibility of the already established exposure limits, and thus lead to an amplification of the risk perception and possibly thus contribute to public outrage.

Wiedemann and Schütz [82] as well as Wiedemann *et al.* [83] conducted a series of experiments with the goal of evaluating whether a policy of precaution taking is effective in decreasing risk perceptions of mobile telephony. They focused on four precautionary policies: stricter limits, protection of sensitive locations, exposure minimization and stakeholder participation in base station siting.

The first experiment was conducted in Austria, the second experiment in Switzerland. Both experiments were comparably designed. In both experiments, the subjects receiving information about precautionary measures expressed a higher risk perception than subjects not receiving such information (Figure 14.2). These differences were statically significant in both studies.

The findings showed that information about implemented precautionary measures does not necessarily decrease risk perceptions. Quite the opposite, the experimental results indicate that precautionary actions tend to amplify risk perceptions, presumably because they are perceived as indicators for a risk. Similar results have recently been found by Barnett and Timotijevic [84] in a UK study on public responses to precautionary information on possible health risks from mobile phones.

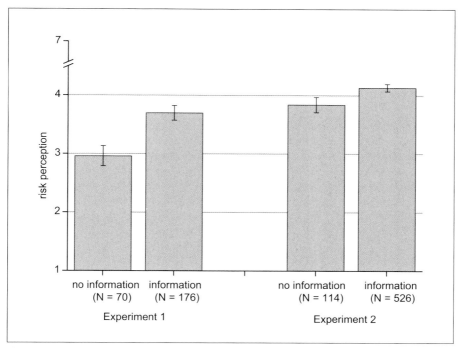

Figure 14.2 Mean ratings (and standard errors of mean) of perceived risk for those receiving information about precautionary measures and those not receiving information about precautionary measures.

Information about precautionary measures may have an amplifying impact on risk perception. In other words, if individuals are informed about the implementation of precautionary measures, they perceive the risks as being greater than those who were not informed about the precautionary measures.

14.5
Outlook and Conclusions

Communicating information on risk assessments and the underlying evidence as well as about remaining uncertainties is a challenging task. Unfortunately, risk communication research can only offer little help because there are considerable knowledge gaps. Only recently have studies started to shed light on the issue of how lay people evaluate scientific studies that underlie risk assessments. Therefore, only a few evidence-based statements can be made.

First of all, it seems that nonexperts have only a limited understanding of the risk assessment framework. They reveal several misperceptions and reach incorrect conclusions in risk assessment. Lay persons also tend to be insensitive with respect to base rates which may result in wrong interpretation of information about relative risks.

These basic problems in interpreting risk information are aggravated by lay people's tendency to ignore information that does not support their prior beliefs. The existing research suggests that effects of risk communication depend strongly on the prior beliefs of the recipients. Once a risk belief has been developed, it appears to be extremely difficult to change people's risk perception.

A further issue is uncertainty. Contrary to the popular assumption that revealing uncertainties will strengthen trust into the risk assessor, it seems that the disclosure of uncertainties can also have negative effects. As mentioned before, the available studies support the assumption that nonexperts have difficulties to differentiate among different types of uncertainty and tend to draw inappropriate conclusions for their risk appraisal.

An additional issue is the understanding of verbal uncertainty categories such as "probable" or "possible". Available research clearly indicates that lay people may have quite different interpretations of those categories. Therefore, they should be used with caution for classifying unclear risks. Especially, it seems to be rather improbable that lay people will draw the proper conclusions from these descriptions for unclear risks without any further explanation.

A rather sensitive problem is precaution taking. Information about precautionary measures can have an amplifying impact on risk perception. Research indicates that when individuals are informed about the implementation of precautionary measures, they may perceive the risks as being greater than those people who were not informed about the precautionary measures. Thus, precautionary measures may be taken as confirmatory proof that a risk exists.

Taken together all issues, one can conclude that risk communicators face in general a difficult task. However, translating complex scientific data into everyday language understandable to lay people is especially challenging when scientific evidence is sparse and study results are inconsistent or difficult to interpret. Lay people may come to the wrong conclusions about the available evidence, its strength or the magnitude of the implied risk. The danger of misinterpretation is always present.

Finally, it becomes evident that our current knowledge is limited with respect to the finer details of the risk and uncertainty communication process including how to describe the underlying evidence as well as the remaining uncertainties for a suspected risk factor. Therefore, we need more evidence-based risk communication research (see, e.g. Ref. [63]) and we should study the risk communication process from a broader perspective. In line with work on intuitive toxicology [13] as well as studies by Fox and Irwin [68] on communication of uncertain beliefs, the role that context plays in risk communication should be highlighted and more attention should be given to the underlying cognitive framework that empowers people to make informed risk judgments.

References

1 Levin, R. (2005) *Uncertainty in Risk Assessment – Contents and Modes of Communication*, Division of Philosophy, Department of Philosophy and the History of Technology, Royal Institute of Technology, Stockholm [http://www.infra.kth.se/FIL/arkiv/Kappor/Lic%20Rikard%20Plan%20liss%201%208%20(28%209%20-05)%20p.pdf] [Retrieved: 19.2.2008].

2 Krupnick, A., Morgenstern, R., Batz, M., Nelson, P., Burtraw, D., Shih, J.S. and McWilliams, M. (2006) *Not a Sure Thing: Making Regulatory Choices under Uncertainty*, Resources for the Future, Washington, DC [http://www.rff.org/rff/News/Features/Not-a-Sure-Thing.cfm] [Retrieved: 31.07.2007].

3 Budescu, D.V. and Wallsten, T.S. (1995) Processing linguistic probabilities: General principles and empirical evidence. In *The Psychology of Learning and Motivation: Decision Making from the Perspective of Cognitive Psychology*, (eds J.R. Busemeyer, R. Hastie and D.L. Medin), Academic Press, San Diego, CA, pp. 275–318.

4 National Research Council (1983) *Risk Assessment in the Federal Government: Managing the Process*, National Academy Press, Washington, DC.

5 Finkel, A.M. (1990) *Confronting Uncertainty in Risk Management: A Guide for Decision-makers*, Center for Risk Management, Resources for the Future, Washington, DC.

6 Morgan, M.G. and Henrion, M. (1990) *Uncertainty. A Guide to Dealing with Uncertainty in Quantitative Risk and Policy Research*, Cambridge University Press, Cambridge.

7 National Research Council (1994) *Science and Judgment in Risk Assessment*, National Academy Press, Washington, DC.

8 EPA (2000) *Risk Characterization Handbook*, US Environmental Protection Agency, Washington, DC [http://www.epa.gov/osp/spc/rchandbk.pdf] [Retrieved: 31.08.2007].

9 International Programme On Chemical Safety (2004) *Key Generic Terms used in Chemical Hazard/Risk Assessment*, World Health Organization, Geneva.

10 Wiedemann, P.M., Karger, C. and Clauberg, M. (2002) *Risikofrüherkennung im Bereich Umwelt und Gesundheit. Machbarkeitsstudie für das Aktionsprogramm "Umwelt und Gesundheit"*

im Auftrag des Umweltbundesamtes, Forschungszentrum Jülich, MUT. Umweltforschungsplan des Bundesministeriums für Umwelt, Naturschutz und Reaktorsicherheit, F+E-Vorhaben, 200 61 218/09.

11 Moss, R.H. and Schneider, S.H. (2000) Uncertainties in the IPCC TAR: Recommendations to lead authors for more consistent assessment and reporting. In *Guidance Papers on the Cross Cutting Issues of the Third Assessment Report of the IPCC*, (eds R. Pachauri, T. Taniguchi, and K. Tanaka), World Meteorological Organization, Geneva, pp. 33–51.

12 Lion, R., Meertens, R.M. and Bot, I. (2002) Priorities in information desire about unknown risks, *Risk Analysis*, **22**, 765–776.

13 Kraus, N., Malmfors, T. and Slovic, P. (1992) Intuitive toxicology: Expert and lay judgments of chemical risk, *Risk Analysis*, **12**, 215–232.

14 Neil, N., Malmfors, T. and Slovic, P. (1994) Intuitive toxicology: expert and lay judgments of chemical risks, *Toxicology and Pathology*, **22**, 198–201.

15 Slovic, P., Malmfors, T., Krewski, D., Mertz, C.K., Neil, N. and Bartlett, S. (1995) Intuitive toxicology. II. Expert and lay judgments of chemical risks in Canada, *Risk Analysis*, **15**, 661–675.

16 Slovic, P., Malmfors, T., Mertz, C.K., Neil, N. and Purchase, I.F. (1997) Evaluating chemical risks: results of a survey of the British Toxicology Society, *Human and Experimental Toxicology*, **16**, 289–304.

17 MacGregor, D.G., Slovic, P. and Malmfors, T. (1999) "How exposed is exposed enough?". Lay inferences about chemical exposure, *Risk Analysis*, **19**, 649–659.

18 Ames, B.N., Durston, W.E., Yamasaki, E. and Lee, F.D. (1973) Carcinogens are mutagens: a simple test system combining liver homogenates for activation and bacteria for detection, *Proceedings of the National Academy of the United States of America*, **70**, 2281–2285.

19 Purchase, I.F.H. and Slovic, P. (1999) Quantitative risk assessment breeds fear, *Human and Ecological Risk Assessment*, **5**, 445–453.

20 Gyrd-Hansen, D., Kristiansen, I.S., Nexoe, J. and Nielsen, J.B. (2003) How do individuals apply risk information when choosing among health care interventions? *Risk Analysis*, **23**, 697–704.

21 Halpern, D.F., Blackman, S. and Salzman, B. (1989) Using statistical risk information to assess oral-contraceptive safety, *Applied Cognitive Psychology*, **3**, 251–260.

22 Bar-Hillel, M. (1980) The base-rate fallacy in probability judgments, *Acta Psychologica*, **44**, 211–233.

23 Koehler, J.J. (1996) The base rate fallacy reconsidered: descriptive, normative, and methodological challenges, *Behavioral and Brain Sciences*, **19**, 1–53.

24 Matthews, R.A.J. (1996) Base-rate errors and rain forecasts, *Nature*, **382**, 766–766.

25 Ellsberg, D. (1961) Risk, ambiguity, and the savage axioms, *Quarterly Journal of Economics*, **75**, 643–669.

26 Camerer, C.F. and Weber, M. (1992) Recent developments in modeling preferences – uncertainty and ambiguity, *Journal of Risk and Uncertainty*, **5**, 325–370.

27 Johnson, B.B. and Slovic, P. (1995) Presenting uncertainty in health risk assessment: initial studies of its effects on risk perception and trust, *Risk Analysis*, **15**, 485–494.

28 Johnson, B.B. and Slovic, P. (1998) Lay views on uncertainty in environmental health risk assessment, *Journal of Risk Research*, **1**, 261–279.

29 Johnson, B.B. (2003) Further notes on public response to uncertainty in risks and science, *Risk Analysis*, **23**, 781–789.

30 Kuhn, K.M. (2000) Message format and audience values: Interactive effects of uncertainty information and environmental attitudes on perceived risk, *Journal of Environmental Psychology*, **20**, 41–51.

31 Miles, S. and Frewer, L.J. (2003) Public perception of scientific uncertainty in

relation to food hazards, *Journal of Risk Research*, **6**, 267–283.
32. Ogden, J., Fuks, K., Gardner, M., Johnson, S., McLean, M., Martin, P. and Shah, R. (2002) Doctors expressions of uncertainty and patient confidence, *Patient Education and Counseling*, **48**, 171–176.
33. Hyland, K. (1998) *Hedging in Scientific Research Articles*, John Benjamins, Amsterdam.
34. Levin, R., Hansson, S.O. and Ruden, C. (2004) Indicators of uncertainty in chemical risk assessments, *Regulatory Toxicology and Pharmacology*, **39**, 33–43.
35. Flugstad, A.R. and Windschitl, P.D. (2003) The influence of reasons on interpretations of probability forecasts, *Journal of Behavioral Decision Making*, **16**, 107–126.
36. IARC (2006) *Preamble to the IARC Monographs. IARC Monographs Programme on the Evaluation of Carcinogenic Risks to Humans*, International Agency for Research on Cancer, Lyon [http://monographs. iarc.fr/ENG/Preamble/CurrentPreamble.pdf] [Retrieved: 24.09.2007].
37. Cogliano, V.J., Baan, R.A., Straif, K., Grosse, Y., Secretan, M.B., El Ghissassi, F. and Kleihues, P. (2004) The science and practice of carcinogen identification and evaluation, *Environmental Health Perspectives*, **112**, 1269–1274.
38. Moxey, L.M. and Sanford, A.J. (1993) *Communicating Quantities: A Psychological Perspective*, Lawrence Erlbaum, Hillsdale, NJ.
39. Beyth-Marom, R. (1982) How probable is probable – a numerical translation of verbal probability-expressions, *Journal of Forecasting*, **1**, 257–269.
40. Brun, W. and Teigen, K.H. (1988) Verbal probabilities: ambiguous, context-dependent, or both? *Organizational Behavior and Human Decision Processes*, **41**, 390–404.
41. Clark, D.A. (1990) Verbal uncertainty expressions – a critical review of 2 decades of research, *Current Psychology Research & Reviews*, **9**, 203–235.
42. Weber, E.U. and Hilton, D.J. (1990) Contextual effects in the interpretations of probability words: Perceived base rate and severity of events, *Journal of Experimental Psychology: Human Perception and Performance*, **16**, 781–789.
43. Fillenbaum, S., Wallsten, T.S., Cohen, B.L. and Cox, J.A. (1991) Some effects of vocabulary and communication task on the understanding and use of vague probability-expressions, *American Journal of Psychology*, **104**, 35–60.
44. Fischer, K. and Jungermann, H. (1996) Rarely occurring headaches and rarely occurring blindness: is rarely = rarely? The meaning of verbal frequentistic labels in specific medical contexts, *Journal of Behavioral Decision Making*, **9**, 153–172.
45. Fischer, K. and Jungermann, H. (2003) "Zu Risiken und Nebenwirkungen fragen Sie Ihren Arzt oder Apotheker": Kommunikation von Unsicherheit im medizinischen Kontext, *Zeitschrift für Gesundheitspsychologie*, **11**, 87–98.
46. Jablonowski, M. (1994) Communicating risk: words or numbers? *Risk Management*, **41**, 47–50.
47. Champaud, C. and Bassano, D. (1987) Argumentative and informative functions of French intensity modifiers "presque" (almost), "a peine" (just, barely) and "a peu pres" (about): an experimental study of children and adults, *Cahiers De Psychologie Cognitive/Current Psychology of Cognition*, **7**, 605–631.
48. Teigen, K.H. and Brun, W. (1999) The directionality of verbal probability expressions: Effects on decisions, predictions, and probabilistic reasoning, *Organizational Behavior and Human Decision Processes*, **80**, 155–190.
49. Teigen, K.H. and Brun, W. (2000) Ambiguous probabilities: when does $p = 0.3$ reflect a possibility, and when does it express a doubt? *Journal of Behavioral Decision Making*, **13**, 345–362.

50 Gonzalez, M. and Frenck-Mestre, C. (1993) Determinants of numerical versus verbal probabilities, *Acta Psychologica*, **83**, 33–51.

51 Hamm, R.M. (1991) Selection of verbal probabilities – a solution for some problems of verbal probability expression, *Organizational Behavior and Human Decision Processes*, **48**, 193–223.

52 Patt, A.G. and Schrag, D.P. (2003) Using specific language to describe risk and probability, *Climatic Change*, **61**, 17–30.

53 Thalmann, A.T. (2005) *Risiko Elektrosmog. Wie ist Wissen in der Grauzone zu kommunizieren?* Beltz-Verlag, Weinheim.

54 Neutra, R.R. and DelPizzo, V. (2002) Transparent democratic foresight strategies in the California EMF Program, *Public Health Rep*, **117**, 553–563.

55 Hattis, D. (1989) Scientific uncertainties and how they affect risk communication. In *Effective Risk Communication. The Role and Responsibility of Government and Nongovernment Organizations*, (eds V.T. Covello, D.B. McCallum, and M.T. Pavlova), Plenum Press, New York, pp. 117–126.

56 Magat, W.A., Viscusi, W.K. and Huber, J. (1987) Risk-dollar tradeoffs, risk perceptions, and consumer behaviour. In *Learning About Risk: Consumer and Worker Responses to Hazard Information*, (eds W.K. Viscusi and W.A. Magat), Harvard University Press, Cambridge, MA, pp. 83–97.

57 Reyna, V.F. (1991) Class inclusion, the conjunction fallacy, and other cognitive illusions, *Developmental Review*, **11**, 317–336.

58 Reyna, V.F. and Brainerd, C.J. (1991) Fuzzy-trace theory and framing affects in choice: gist extraction, truncation, and conversion, *Journal of Behavioral Decision Making*, **4**, 249–262.

59 Kahneman, D. and Tversky, A. (1979) Prospect theory – analysis of decision under risk, *Econometrica*, **47**, 263–291.

60 Viscusi, W.K. (1997) Alarmist decision with divergent risk information, *The Economic Journal*, **107**, 1657–1670.

61 Schapira, M.M., Nattinger, A.B. and McHorney, C.A. (2001) Frequency or probability? A qualitative study of risk communication formats used in health care, *Medical Decision Making*, **21**, 459–467.

62 Lipkus, I.M. and Hollands, J.G. (1999) The visual communication of risk, *Journal of the National Cancer Institute Monographs*, 149–163.

63 Trevena, L.J.H.M. Barratt, A., Butow, P. and Caldwell, P., (2006) A systematic review on communicating with patients about evidence, *Journal of Evaluation in Clinical Practice*, **12**, 13–23.

64 Ancker, J.S., Senathirajah, Y., Kukafka, R. and Starren, J.B. (2006) Design features of graphs in health risk communication: a systematic review, *Journal of the American Medical Informatics Association*, **13**, 608–618.

65 Ibrekk, H. and Morgan, M.G. (1978) Graphical communication of uncertain quantities to nontechnical people, *Risk Analysis*, **7**, 519–529.

66 Thompson, K.M. and Bloom, D.L. (2000) Communication of risk assessment information to risk managers, *Journal of Risk Research*, **3**, 333–352.

67 Van Dijk, T.A.(ed.) (1997) *Discourse as Structure and Process. Discourse Studies: A Multidisciplinary Introduction*, **Vol. 2**, Sage, London.

68 Fox, C.R. and Irwin, J.R. (1998) The role of context in the communication of uncertain beliefs, *Basic and Applied Social Psychology*, **20**, 57–70.

69 Ito, T.A., Larsen, J.T., Smith, N. and Cacioppo, J.T. (1998) Negative information weighs more heavily on the brain: the negativity bias in evaluative categorizations. *Journal of Personality Social Psychology*, **75**, 887–900.

70 Taylor, S.E. (1991) Asymmetrical effects of positive and negative events – The mobilization minimization hypothesis, *Psychological Bulletin*, **110**, 67–85.

71 Koehler, J.J. (1993) The influence of prior beliefs on scientific judgments of evidence

quality, *Organizational Behavior and Human Decision Processes*, **56**, 28–55.

72 Evans, J. (1990) *Bias in Human Reasoning*, Lawrence Erlbaum, Hillsdale, NJ.

73 Wiedemann, P.M., Schütz, H., Thalmann, A. and Grutsch, M. (2006) Mobile fears? – Risk perceptions regarding RF EMF. In *Risk Perception and Risk Communication: Tools, Experiences and Strategies in Electromagnetic Fields Exposure*, (eds C. del Pozo, D. Papameletiou, P.M. Wiedemann and P. Ravazzani), pp. 35–46, Servizio Pubblicazioni e Informazioni Scientifiche, Rome.

74 Smithson M., (1999) Conflict aversion: preference for ambiguity vs conflict in sources and evidence, *Organizational Behavior and Human Decision Processes*, **79**, 179–198.

75 Hertwig, R., Barron, G., Weber, E.U. and Erev, I. (2004) Decisions from experience and the effect of rare events in risky choice, *Psychological Science*, **15**, 534–539.

76 Earle, T.C., Siegrist, M. and Gutscher, H. (2006) Trust, risk perception and the TCC model of cooperation. In *Trust in Cooperative Risk Management*, (eds M. Siegrist, T.C. Earle, and H. Gutscher), Earthscan, London, pp. 1–49.

77 Hunt, S. and Frewer, L. (2001) Trust in sources of information about genetically modified food risks in the UK, *British Food Journal*, **103**, 46–62.

78 White, M.P., Pahl, S., Buehner, M. and Haye, A. (2003) Trust in risky messages: the role of prior attitudes, *Risk Analysis*, **23**, 717–726.

79 Zukier, H. and Pepitone, A. (1984) Social roles and strategies in prediction – some determinants of the use of base-rate information, *Journal of Personality and Social Psychology*, **47**, 349–360.

80 Schwarz, N., Strack, F., Hilton, D. and Naderer, G. (1991) Base rates, representativeness, and the logic of conversation – the contextual relevance of irrelevant information, *Social Cognition*, **9**, 67–84.

81 Dennis, M.R. and Babrow, A.S. (2005) Effects of narrative and paradigmatic judgmental orientations on the use of qualitative and quantitative evidence in health-related inference, *Journal of Applied Communication Research*, **33**, 328–347.

82 Wiedemann, P.M. and Schütz, H. (2005) The precautionary principle and risk perception: experimental studies in the EMF area, *Environmental Health Perspectives*, **113**, 402–405.

83 Wiedemann, P.M., Thalmann, A.T., Grutsch, M.A. and Schütz, H. (2006) The impacts of precautionary measures and the disclosure of scientific uncertainty on EMF risk perception and trust, *Journal of Risk Research*, **9**, 361–372.

84 Barnett, J., Timotijevic, L., Shepherd, R. and Senior, V. (2007) Public responses to precautionary information from the Department of Health (UK) about possible health risks from mobile phones, *Health Policy*, **82**, 240–250.

15
Ethical Guidance for Dealing with Unclear Risk
Armin Grunwald

15.1
Ethical Guidance in Cases of Unclear Risk – The Challenge

Risk assessment and risk management strategies and approaches have been developed and established in many areas of science, medicine and technology [1, 2], e.g. in dealing with new chemicals or pharmaceuticals. Such "classical" risk regulation is adequate if the level of protection is defined and if the risk can be quantified, e.g. as a product of the probability of occurrence of the adverse effects and the assumed extent of possible damage, or by using other quantifying schemes like: risk denotes "the probability of an adverse effect in an organism, system, or (sub)population caused under specified circumstances by exposure to an agent" [3]. In such situations, thresholds can be set by law, by self-commitments or by applying participatory procedures and measures can be taken to keep particular effects well below predefined thresholds [4].

However, in many cases the conditions for such classical risk regulation are not or not sufficiently fulfilled (see Section 15.2.2). There may be situations of an "inconclusive risk assessment" or of "unclear risk": it might not be clear whether the situation under consideration could involve risks at all or that the extent of possible damage might not be known (a perfect illustration of the problems and open questions related to such a situation of "unclear risk" is provided by the present debate on possible harmful effects of nanoparticles [5, 6]). Reliable knowledge concerning possible adverse effects might not be available yet or might be persistently controversial and hypothetical. Empirical evidence could still be missing and data or their interpretation might be conflicting.

In such cases, often rigid postulates are formulated. A recent example was the postulate of the ETC group for a moratorium in the case of nanoparticles [7]:

> At this stage, we know practically nothing about the possible cumulative impact of human-made nano-scale particles on human health and the environment. Given the concerns raised over nano-particle contamination in living organisms, the ETC group proposes that governments declare an

> immediate moratorium on commercial production of new nano-materials
> and launch a transparent global process for evaluating the socio-economic,
> health and environmental implications of the technology

This statement caused fears and concerns on the side of the nano-scientists that public rejection of nanotechnology might be the consequence (which, however, did not happen yet). This case shows the high degree of irritation and the difficulties of finding a "rational" way for dealing with "unclear risk". Especially in situations where decisions also have to be made in such unclear cases, guidance and orientation are needed. The challenge is then to support the respective decision-making situations by providing guidance in order to prevent decisions from being arbitrary or the subject of mere ideology.

Therefore, guidance in questions of this type is urgently needed. In this chapter, we will concentrate on what *ethical* reflection and guidance could contribute to dealing with such situations. To this end it is necessary first to identify the entry points for ethical reflection which are or might be involved in situations of unclear risk or of high uncertainty (Section 15.2). Second, selected approaches of philosophical ethics on how to deal with these situations will be introduced and assessed with respect to their relevance for problem solving (Section 15.3). Third, the "precautionary principle" and the "prudent avoidance approach" as operative approaches to the problem of decision making under high uncertainty will be discussed with respect to their ethical foundation (Section 15.4). The conclusions (Section 15.5) are dedicated to a brief assessment of the chances and limitations of what might be expected from ethical guidance in the situations considered.

15.2
Entry Points of Ethical Reflection in Situations of Unclear Risk

There are different meanings of the notion of "ethics" among different scientific disciplines, societal groups, policymakers and also among different philosophical traditions. Therefore some clarifying remarks on the meaning of "ethics" applied in this chapter will be given (Section 15.2.1). Second, the important notions of unclear risk and uncertainty have to be clarified, too (Section 15.2.2). The third task of this section is identifying the specific types of moral conflicts that might emerge in situations of unclear risk (Section 15.2.3) – such moral conflicts are, due to the applied understanding of "ethics", the "entry points" for possible ethical guidance.

15.2.1
Entry Points of Ethical Reflection in General

In the modern discussion the distinction between factual morals, on the one hand, and ethics as the reflective discipline in cases of moral conflicts or ambiguities, on the other hand, has been widely accepted [8, 9]. This distinction takes into account the plurality of morals in modern society. As long as established traditional moral convictions (e.g. religious ones) are uncontroversial and valid among all relevant actors, and as long as they are sufficient to deal with the respective situation and do not leave open relevant questions, ethical reflection is not in place [10, 11]. Morals are, in fact, the action-guiding maxims and rules of an individual, of a group or of society as a whole which can be analyzed empirically by the social sciences.

Ethical analysis takes, on the other hand, these morals as subjects to reflect about in normative respect. Ethics is concerned with the *justification* of moral rules of action at the normative level, allowing better-founded requests beyond the respective, merely particular morals. In particular, ethics serves the resolution of conflict situations which result out of the actions or plans of actors based on divergent moral conceptions purely by argumentative deliberation. In the history of philosophy some key approaches have been developed like the Categorical Imperative proposed by Kant or the idea of the Pursuit of Happiness formulated by Bentham (Utilitarianism).

Suspicions about unclear risks lead, in an open and morally pluralistic society, unavoidably to societal debates at the least and often also to conflicts about their acceptability. We could witness recent examples in the fields of nuclear power and radioactive waste disposal, stem cell research, genetically modified organisms, and reproductive cloning. Those conflicts are often controversies about specific *futures*: expectations and desires, fears and hopes which are uncertain and, as a rule, highly contested [12]. The role of ethics consists of the analysis of the normative structure of conflicts over risks, and of the search for rational, argumentative, and discursive methods of resolving them (in this "continental" understanding ethics is primarily part of the philosophical profession but might be related with participative approaches [13] or might be part of broad interdisciplinary approaches [14]). In ethical reflection in the various areas of application, however, there are close interfaces to and inevitable necessities for interdisciplinary cooperation with the natural and engineering sciences involved as well as with the humanities.

15.2.2
Unclear Risk: Nonstandard Situations with Respect to Risk

In order to explain the notion of "unclear risk" we will generally consider a decision-making situation or the respective preparatory or preceding debate where knowledge about possible harmful effects has to be taken into account. The *differentia specifica* for clarifying the notion of unclear risk compared to "standard" risk will be the availability and validity of the information about possible harmful effects (following Ref. [4]).

"Standard" risk is a *quantified magnitude* denoting the probability of the occurrence of the adverse effects multiplied by their impact. In situations with such quantitative risk measures available policy makers can respond with a classical risk management approach whereby, for example, thresholds can be set and risks can be either minimized or kept below a certain level according to the respective level of protection. As there is *scientifically sound and consolidated* knowledge concerning the adverse effects in question – with respect to the probability of their occurrence as well as concerning the expectable damage – society can act by applying familiar preventive interventions (e.g. by building higher dams in case of expected higher floods). Laws like the Toxic Substance Control Act in the US [15] determine what has to be done in such standard risk situations. Consequently, ethical reflection is not required and the problem under consideration may be approached directly using the established risk regulation measures.

While "standard" risk is regarded as a quantifiable parameter, "unclear" risk is denoting a "nonstandard situation" in moral respect [11] with respect to risk which can be characterized by limits to quantifiability, by lack of knowledge and data, by conflicting data or by conflicts about their interpretation, by epistemic uncertainties and/or by unsolved scientific controversies [4]. The following cases can be distinguished analytically:

(1) The most relevant case in this respect is simply *lack of knowledge* about cause–effect relations that does not allow us to answer the question whether there might be adverse effects at all. Concerning the far-reaching difference between *possible* and *probable* effects, this case covers the situation of debating about *merely possible* adverse effects. In this case we could neither calculate the probability of the occurrence of adverse effects nor the magnitude of possible damage. It would be a typical situation of unclear risk. The current debate about possible harmful effects of nanoparticles on human health or the environment may again be taken as an example [5, 6].

(2) Classical risk approaches are limited with respect to their applicability when particular cause–effect relationships have not yet been scientifically established while at the same time the adverse effects can already be observed. The case of bovine spongiform encephalopathy was of this nature until scientific evidence about the cause–effect chain leading to the "mad cow" phenomenon had been established. In such cases there are visible harmful effects, but it is neither possible to identify the main origins of those effects nor to calculate the probability of their occurrence.

(3) The risk approach might also be difficult to apply if we cannot fully rely on the scientific information estimating possible adverse effects. This is notably the case when an epistemic debate is going on in science: different disciplines use competing models or analogies or basic assumptions to disclose the subject matter under investigation in order to acquire new knowledge. According to deep-reaching contradictions in the basic scientific conceptualizations there will be an unclear situation concerning the data available and their interpretation [*cf.* [4] for the case of genetically modified organisms (GMOs)].

Decision making in these cases is obviously different from a classical risk assessment situation: we will denote these types of situations as decision making with "unclear risk". (Frequently, the notion of uncertainty is used for these cases, and "risk" and "uncertainty" form the basic distinction. However, this notion covers the fact that also quantifiable risk is, in a certain sense "uncertain" because of the preliminary nature and incompleteness of knowledge. In the notion to be used throughout this chapter "risk" and "unclear risk" both are uncertain, but showing a different degree of uncertainty.)

The question arises whether "unclear risk" in this sense can be distinguished, on the other side, from mere speculation. If there is, in the first case mentioned above, no knowledge at all about possible harmful effects, suspicions concerning unclear risk would be purely speculative, with all the imaginable consequences for decision making. Therefore, it should be part of the questions to be addressed whether (even weak) arguments for expecting adverse effects could be given with some "evidence" or plausibility of the assumptions under consideration, but without having full scientific evidence. There will be a need not only to distinguish between quantified risk and unclear risk, but also between unclear risk and mere speculation.

15.2.3
Moral Conflicts in Situations of Unclear Risk

The next task is to identify points in decision-making situations involving unclear risk which might be subject to ethical inquiry according to the understanding of ethics given above. Such situations form "nonstandard" situations in moral respect (according to the notion in Ref. [11]) because the normative framework (here, the system of risk regulation in place) governing the situation is not able to allow a conclusive statement and clear indication for corresponding action. In this indifferent or ambiguous situation we have to look for other types of orientation. Ethical reflection is required, according to Section 15.2.1, in the case of moral conflicts. There are, indeed, some typical questions in this respect:

- *Acceptability of unclear risk.* The most general ethically relevant question is whether and under which preconditions we should tolerate unclear risk, and what criteria should govern the question of acceptability. Strong precautionary approaches could motivate the postulate to avoid completely unclear risks. In questions of technology, for example, this would imply not introducing any new material, product or system without having full knowledge about possible harmful consequences and impacts (*cf.* [7]). Jonas restricted this strong precautionary approach to situations with potentially apocalyptic character (see Section 15.3.3). On the other side, there are more or less hidden "wait and see" strategies [16] taking nonknowledge about adverse effects as permission. It is easy to see that different moral positions stand behind those differing approaches to the problem of unclear risk. Insofar as moral conflicts may be expected with respect to unclear risk (and this has been the case, for example, in the case of GMOs [4]) where ethical inquiry is required.

- *Weighing benefits against unclear risks.* Often, there are clear and well-founded assumptions about expected benefits by developing specific technologies further. As an example, a huge market potential is usually ascribed to nanoparticles and nanomaterials. The ethical question then arises as to what role such *opportunities* may play in decision-making situations involving unclear risk. Is it ethically legitimate to proceeding further with technology development and implementation under consideration in the presence of large knowledge gaps or of scientific controversies about risk? Two problems become visible at this point. (i) The principal question whether weighing benefits against unclear risks would be justified at all as a decision-making approach – there might be ethical knockout arguments against an unlimited permission of weighing pros and cons, e.g. in cases with a potential of an extremely high damage up to a catastrophe for humankind (Section 15.3.3). (ii) There is a problem at the methodical level: how could we perform such a balancing in the absence of quantitative measures, at least at the side of possible risk? At the heart of ethical inquiry is the question – which frequently leads to conflicts – upon which criteria (often well-known) benefits and (in the situation of uncertainty unclear) risks could be weighed against each other allowing a rational decision-making process.

- *Normalizing the situation under consideration.* A familiar approach to provide orientation is to propose analogies of the new situation to other situations we are familiar with. A new (and perhaps irritating) decision-making situation might be "normalized" be referring to a well-known situation. This approach, however, depends on the acceptance of some premises. Building an analogy is not a value-neutral step because it always presupposes the similarity of the new to older situations. Assessing the similarity, however, depends on the determination of criteria of similarity and this choice is a normative step in building analogies. For example, the promotors of GMOs constructed an analogy of the genetic modification of organisms in laboratories to traditional approaches in agriculture. Both approaches aim at changing the genome of the affected organisms in a desired direction. The opponents, however, pointed to differences in other respects, e.g. concerning the timespan of modifications, and rejected this analogy. Therefore, the choice of criteria of building analogies might cause conflicts and is therefore a candidate for ethical reflection in cases of uncertainty. The question is whether comparisons of unclear risks to other types of unclear risk would be justified. For example, a question with moral conflict potential could be whether criteria for assessing nanoparticle risks could be transferred from experiences in the field of new chemicals or pharmaceuticals.

- *Comparisons of man-made situations of unclear risk with a natural background.* The normalization of new and irritating situations is sometimes attempted by pointing to natural exposure. A well-known example from the field of "clear" risk is the attempt to normalize the exposition to radioactive radiation by pointing to the fact that we are exposed to natural radioactivity, in some regions to a considerable dose. A problem similar to the previous point arises at this point. Comparisons are valid only relative to accepted criteria. At least one criterion is not considered in this example mentioned above: the situation that exposure to natural radiation is really a "natural" hazard and therefore external to human influence while radioactive radiation from technical installations is man-made. Therefore, many concerned people did not accept the normative implications of that comparison that they should accept the artificial exposure because of the pre-existing and, in quantitative terms, larger natural exposure. The case shows the conflict potential at the level of normative criteria with the consequence of ethical reflection might be required in arguing pro or contra the comparability of man-made to natural exposure.

- *Learning from historic cases.* Some voices point to the analogy of artificial nanoparticles created in industry to asbestos: "Some people have asked whether the ultra-small particles and fibers that nanotechnology produces, such as carbon nanotubes, might become the new asbestos" [17]. Could we learn something from the asbestos story [18] or from other stories [19] and under which conditions could we transfer that knowledge to newly emerging cases like the nanoparticle case? Ethics again would look here for the normative criteria of transferring knowledge and building analogies.

Ethics' contribution to this subject is, above all, an analytical one [11]. It consists of a value judgment of the situation (values involved, possible moral conflicts, applicability of the precautionary principle), in a clarification of the comparability of new unclear risks to other types of risk, and in disclosing the normative presuppositions and implications entering into it, as well as in investigating the normative basis for practical consequences. These questions of the acceptability and comparability of risks, the advisability of weighing up risks against opportunities, and the rationality of action under uncertainty are at the heart of ethical reflection in the case of unclear risk.

15.3
Ethical Approaches to (Unclear) Risk

There is a lot of ethical work available in the context of risk in general [2, 20, 21], but only few activities specifically on unclear risk. In the following we will describe and assess the consequentialist approach and its limits (Section 15.3.1), the principle of pragmatic consistency (Section 15.3.2), the ethics of responsibility (Section 15.3.3), a recent approach on "projected time" (Section 15.3.4) and deontological advice (Section 15.3.5).

15.3.1
Consequentialist Approach

Consequential ethics focuses on the expected consequences and impacts of actions and decisions rather than on the intentions standing behind them. Insofar as the occurrence of unclear risk would be a consequence of actions or decisions, a consequentialist approach would be the ideal point of departure in cases of ethical reflection being required (according to the types of questions mentioned above). Even in the case of a Kantian approach – which would look primarily on the intentions of action rather than on expected consequences – it would be unavoidable to take into account also assumed consequences in order to get a comprehensive picture of the situation under consideration [8, 22]. In a certain sense, ethical reflection of what should be done or not done is closely related with considering consequences of actions and decisions anyway, independent of the specific ethical approach. In our case, an action or decision would create a situation involving unclear risk which itself would be a consequence – and possibly a risk in its own [23].

However, the notion of "consequentialism" is mostly used in a more narrow sense [21, 23], combining the assessment of consequences of action with a utilitarian approach concerning a specific way of ethical inquiry and judgment. The aim of the utilitarian consequentialist approach is to investigate all relevant outcomes and impacts of a decision in advance – the expected benefits as well as the possible negative effects. This investigation addresses the probability of those outcomes and impacts as well as their extent. Ideally this investigation would lead to relating specific actions or decisions with expected costs and benefits in monetary terms. Ethical analysis would be transformed into balancing risks and benefits using a quantitative calculus [21, 22]. By aggregating these data the utility of the considered options of decision making can be calculated and the option showing the highest expectable utility can be identified, as in the cost–benefit analysis. Often the multi-criteria decision-making approach (MCDA) [24] is used, which consists of an operationalization of the utilitarian principle of maximizing the utility. Risks are taken into account by denoting possible damage as negative utility.

In the case of unclear risk, however, a basic precondition of this approach, the availability of quantitative data on probability and extent of possible damage, is not fulfilled by definition (see above). The consequentialist approach in the narrow sense with its close relation to an economic assessment is methodologically analogous to the classical risk management approach, but does not work in the case of unclear risk. In the absence of accepted procedures of quantification, the consequentialist approach is not applicable (in the same direction, *cf.* [23, 25]). Although this approach has an ethical foundation in utilitarian ethics there is no way to expect guidance in the case of unclear risk due to the methodological difficulties.

15.3.2
Principle of Pragmatic Consistency

Ethical guidance in cases of dealing with risk could consist of relating new situations of risk to risk situations we are familiar with. In this case we could – under specific preconditions (*cf.* Section 15.2.3) transfer our experiences from the established risk strategies to newly emerging situations.

In this field a "principle of pragmatic consistency" has been proposed [26]. It takes as its point of departure the observation that the idea of rationally acting persons is related to the request for consistency. We would have problems to ascribing the attribute "rational" to persons who act in an obviously inconsistent way. The "principle of pragmatic consistency" transfers this observation to the risk problem, postulating that the factual behavior of people in existing situations of risk (e.g. in their lifeworlds) could be taken as measure of what risk types and extents could rationally be assumed to be acceptable for them. By this argument, a link between the empirical risk behavior and normative questions of the acceptability of risks shall be established.

However, there are a lot of problems related to this approach even in cases of "standard situations" with respect to risk [27]. In case of unclear risk, the specific problem arises that the comparison that would have to be made between the existing (and established) risks and the new one would not be possible because of the lack of quantifiable date (see above). Due to the same methodological argument as in Section 15.3.1, this approach is not able to provide guidance in the case of unclear risk.

One could try, however, to generalize the basic idea of this principle – deriving conclusions for the acceptability of risks by combining knowledge about empirical behavior of people and normative standards – to cases of unclear risk. It seems to be promising to have a look at the probably many unclear risks which we all are dealing with in our everyday life and to ask what this situation would or could imply for newly emerging types of unclear risk with regard to normative standards of rationality. In spite of the foreseeable difficulties in making such a principle work this thought would provide an interesting relation between the "Is" (empirical behavior with respect to risk acceptance) and the "Ought" (normative standards of rational behavior) which should be explored in more detail.

15.3.3
"Imperative of Responsibility" (Jonas)

Regarding the fact that both approaches mentioned above need quantified risk values to be applied – which are not available in situations of unclear risk by definition – the question is whether there are ethical approaches available aiming at exactly this situation. The proposals presented in this and the next section depart from the diagnosis that building a product of the probability of the occurrence of harmful effects and the extent of expected damage might, independent from its methodological problems, be problematic in ethical regards. The famous work of Hans Jonas [28] started with the observation that particular technical developments might have an apocalyptic potential threatening the future existence of mankind. According to his normative presupposition "that mankind should be" and that the existence of mankind must not be endangered, Jonas formulated a new and in his option "categoric" imperative: "act in a way that the consequences of your actions are compatible to the permanence of real human life on Earth". These thoughts had an impact, amongst others, for the debate on sustainable development [29].

Coming back to the situation of unclear risk it seems at the first glance that this imperative could provide ethical guidance. Especially in cases with apocalyptic potential in the sense that a danger for the further existence of humankind would be among the unclear risks, Jonas' position seems to give orientation in the direction: stop developing or implementing technology if such situations could be the result. To make his principle more operable Jonas postulated a "heuristics of fear" in order to get an impression of possible negative developments and an obligation to use the worst scenario as orientation for action. If the worst scenario would show an apocalyptic potential, then the respective action should not be taken.

However, many criticisms have been brought up against Jonas' approach. In particular, the naturalistic premises and the supposed teleology of nature used in the derivation of his central imperative, but also the arbitrariness of conclusions have been major arguments. It might be possible in nearly every case to construct a worst scenario showing an apocalyptic dimension. As a consequence, no action at all would ethically be justified – but doing nothing might also cause situations with unclear risk. An aporetic situation could be the consequence, demonstrating that the "imperative of responsibility" does not provide applicable and operable guidance.

The reason for this diagnosis is that Jonas did not formulate any requirements concerning the evidence or probability of the worst-case scenarios. Such scenarios play a decisive role in his argumentation because they determine whether there might be an apocalyptic potential of the respective action or not. For Jonas the mere thinkability of the worst case is sufficient to use those scenarios in an argumentation possible rejecting a whole line of technology. This is exactly the weak point of his approach because this "low level" opens up space for arbitrary speculation about worst cases and then this arbitrariness leads to the aporetic situation outlined above. In this way it becomes clear that Jonas' approach might be very appropriate to raise awareness with regard to situations involving unclear risk. It is, however, completely inadequate as ethical guidance as well as in regulatory debates.

15.3.4
Projected Time

The risk debate on nanotechnology – where most of the risks are "unclear risks" in the sense of this chapter [5, 6] – motivated some authors to think even more radically about ethics in the case of unclear risk. Dupuy and Grinbaum [30] even went beyond Hans Jonas' "Heuristics of Fear" (see above) and formulated a "duty to expect the catastrophe" in order to prevent the catastrophe. The uncertainty of our knowledge about nanotechnology and the consequences in connection with the immense potential for damage, of possibly catastrophic effects (e.g. by losing control over self-replicating robots [31]) are taken as an occasion for categorizing even the precautionary principle (Section 15.4.1) as insufficient. The authors put forward the following argumentation in an approach of "projected time", which makes use of strong assumptions about the future:

- Nobody can know anything about the future of nanotechnology – except that it is the utter catastrophe;
- if everyone could be convinced that nanotechnology is *the* catastrophe, there could be a general renunciation of nanotechnology, so that finally;
- the catastrophe could still be avoided.

This argumentation is paradoxical: if everyone would believe in the assertion that nanotechnology is simply "the catastrophe" and then "renounce nanotechnology", then the catastrophe would not happen, even if it is at present claimed to be certain that the catastrophe will happen. To put it concisely: the catastrophe will not happen because everybody's convinced that it is certain that it will happen. The assumption of the inevitability of the catastrophe has no "validity" in the sense of a discourse between opponents and proponents, but only serves didactic purposes, to motivate a final "renunciation". In this thinking projection of time makes use of the well-known mechanisms of self-fulfilling or self-destroying prophecies: "The predictor, *knowing that his prediction is going to produce causal effects in the world*, must take account of this fact if he wants the future to confirm what he foretold" [30]. The assumed projection, however, fails because of two reasons. (i) Using predictions in order to intentionally influence the further course of development (here, motivate renunciation of nanotechnology in order to prevent the catastrophe) is highly risky. Projection in the sense of a determination of the future runs into the same problems as usual planning approaches. There will be no guarantee to avoid the catastrophe. (ii) The whole argumentation relies on the premise that nanotechnology is "the" catastrophe. Although some arguments are given in favor of this thesis [31], it is highly improbable that most people will not be convinced. However, the whole argumentation chain then fails. Due to these two reasons and assuming that in most cases of unclear risk there will be no overall agreement about something being "the" catastrophe (instead, related futures will usually be contested [12]) the proposed model of "projected time" will not be able to provide ethical guidance.

15.3.5
Deontological Advice

Deontological ethics does not primarily look at the consequences and impacts of actions and decisions, but on human rights and duties, based, for example, on a Kantian approach. Concerning risk, Nida-Rümelin [25] proposed an approach that questions and partly rejects the consequentialist approach in the utilitarian sense (Section 15.3.1). Following this position there are ethical presuppositions hidden in the decision to follow a consequentialist approach, respectively weighing and balancing risks and benefits. Such presuppositions address the question whether a balancing approach would be ethically justified at all. The deontological position [25] emphasizes that there might be strong arguments in favor of or against a particular decision or action which would make a balancing procedure to a morally problematic or even immoral approach. Take, for example, the situation that benefits for a lot of people could only be realized if some people would be victims, concerning health, welfare or even life. In a pure balancing approach applying the principle of maximizing utility it could be the result that it would be allowed that some people should die to realize benefits for the large number of other people (for a famous example, see Ref. [32]).

This type of argumentation leads to the conclusion that there are limits to risk management in the consequentialist, in particular utilitarian, approach (as described in Section 15.3.1). Such limits are given by (following Ref. [25]):

- Human right for life
- Human rights in general (following, for example, the United Nations Declaration of Human Rights)
- Civil rights (e.g. for participation)
- Rights of nonhuman beings like animals (as far as agreed upon due to ethical positions and legal codifications)
- Rights of possession (weaker than the others, but established in modern societies)

Such rights lead, following the deontological approach, to restrictions with regard to the approach of balancing and weighing risks against expected benefits. The deontological position makes the point that there might be ethical problems already in the pre-phase of a risk management procedure. Instead of simply applying a balancing approach, the people affected should first assess whether such a procedure would be ethically legitimate. However, the utility of this statement in cases of unclear risk seems limited: unclear risk exactly concerns the situation that classical risk management is not applicable for methodological and data reasons (Section 15.2.2). Therefore, we are by definition outside of the balancing approach.

What the deontological position could contribute to unclear risk is the hint to the normative dimension of rights and duties. In case of unclear risk there might be situations where a decision causing an unclear risk even might be ethically problematic merely because of the fact that the unclear situation might bring up threats to human rights which would not be the case by making alternative decisions. In such a situation it could be ethically prudent to follow this argumentation. Mostly, however, situations will be more complex. Unclear risks of different types might be the consequence of any of the possible alternatives. In this case the deontological position does not provide orientation because it lacks an operable procedure of comparing different types of unclear risk. It might lead, in these situations, to the same problem of apories like we could observe in Jonas' approach (Section 15.3.3).

15.3.6
Interim Conclusions (1)

The ethical approaches presented above do not provide a clear guidance in situations involving unclear risk. Some of them (utilitarian approach, principle of pragmatic consistency) need quantified risk values so that they are, by definition, applicable only in standard risk situations. Others will lead to aporetic situations (Jonas, Dupuy) because they do not offer an operable strategy of distinguishing between unclear risk and mere speculation. The deontological position gives some hints of what has to be taken into account, but misses the specific challenge of unclear risk.

This conclusion might sound disappointing, but it is not. It would be an exaggerated desire to expect ethics to be able to solve the problem of decision making involving unclear risks by its own means solely. Instead, we will look in the following for ethical support to problem-solving strategies, but without expecting the solution from ethical guidance alone. To this end, we can take the following offers from the above-mentioned approaches for using them in broader problem-solving strategies:

- Ethical approaches indicate which problem dimensions have to be taken into account (e.g. Jonas: apocalyptic potential; deontology: human rights). In this way they contribute to a comprehensive picture of the normative issues touched.

- They foster thinking about alternatives and making comparisons by assessing the alternatives in normative respect as far as possible. The comparative approach

(although being not able to quantify the parameters to be compared in case of unclear risk) allows proceeding in a better structured way.
- Ethical approaches raise the question for a "normalization" of new risk situations by comparing them to well-known situations and include reflection of the criteria to be applied for comparisons.

In this way, ethical guidance contributes to a more comprehensive picture, allows for structuring the normative issues involved and offers analytical work concerning the structure of the respective problem in a normative regard. The aim of the remainder of this chapter is to analyze in what way these contributions could be a fruitful and/or necessary part of broader approaches.

15.4
Operative Approaches

As we can see, ethical approaches to the challenge of unclear risk offer reflective capacities, but do not solve the ambiguity and disorientation problems. In particular, they lack operable strategies for distinguishing between unclear risk and mere speculation. There are two approaches which have been designed to be used operatively in societal decision making – the precautionary principle (Section 15.4.1) and the principle of prudent avoidance (Section 15.4.2). Both of them make explicit or implicit reference to ethical positions.

15.4.1
Precautionary Principle

The observation that in many cases severe adverse effects in the course of the introduction of new materials had not been detected in an early stage but rather led to immense damage on human health, on the environment and also on economy (see impressive case studies in Ref. [19]) motivated debates about precautionary regulation measures. A wide international agreement on the precautionary principle was reached during the Earth Summit (United Nations Conference on Environment and Development) in Rio de Janeiro 1992 and became part of Agenda 21 (as stated by principle 15 of the Rio Declaration):

> In order to protect the environment, the precautionary approach should be widely applied by States according to their capabilities. Where there are threats of serious or irreversible damage, lack of full scientific certainty shall not be used as a reason for postponing cost-effective measures to prevent environmental degradation.

The precautionary principle has been incorporated in 1992 in the Treaty on the European Union: Article 174 postulates:

> Community policy on the environment shall aim at a high level of protection taking into account the diversity of situations in the various regions of the Community. It shall be based on the precautionary principle . . .

The precautionary principle thus establishes a rationale for political action: it substantially lowers the (threshold) level for action of governments (see Ref. [4] for the following). It considerably changes the situation compared to the previous context in which politicians could use (or abuse) a persistent dissent among scientists as a reason (or excuse) simply not to take action at all. In cases for which the accumulation of relevant scientific evidence can take decades, this implies that political action could always be postponed with the argument that scientific knowledge would still have to be completed. In this way, political action could simply come much too late.

It is, however, a difficult task to make legitimate decisions about precautionary measures without either running into the possible high risks of a "wait-and-see" strategy [16] or overstressing precautionary argumentation with the consequence of no longer being able to act (Section 15.3.3.). The following characterization of the precautionary principle shows – in spite of the fact that it still does not cover all relevant aspects – the complex inherent structure of the precautionary principle [4 (modified), 5]:

> Where, following an assessment of available scientific information, there is reasonable concern for the possibility of adverse effects but scientific uncertainty persists, measures based on the precautionary principle may be adopted, pending further scientific information for a more comprehensive risk assessment, without having to wait until the reality and seriousness of those adverse effects become fully apparent.

Thinking about the application of the precautionary principle therefore generally starts with a *scientific examination*.

There is a need to have an assessment of the state of the knowledge available in science, and of the types and extents of uncertainties involved. Drawing the borderline with classical risk management practice or the situation of a purely conjectural risk involves making normative choices that need to be made explicit [20]. In assessing the uncertainties involved, *normative* qualifiers come into play while applying the precautionary principle [4]. It has to be clarified whether there is "reasonable concern" in this situation of uncertainty. The qualifier "reasonable concern" as employed by the European Community guidelines relates to a judgment on the quality of the available information [4]. Therefore, the assessment of the knowledge available including its uncertainties enters the center of precautionary reflections (for methodical proposals, see Ref. [33]). The question of normative guidance is transformed into a procedural challenge of assessing the status of the knowledge available.

15.4.2
Principle of Prudent Avoidance

The prudent avoidance principle states that reasonable efforts to minimize unclear risks should be taken when the actual magnitude of the risks is unknown [34, 35]. The principle was proposed in the context of electromagnetic radiation safety (in particular, fields produced by power lines). A report for the Office of Technology

Assessment of the US Congress [36] described prudent avoidance of power line fields as:

> ... looking systematically for strategies which can keep people out of 60 Hz fields arising from all sources but only adopt those which look to be "prudent" investments given their cost and our current level of scientific understanding about possible risks.

The principle has been adopted in a number of countries, e.g. Sweden, Denmark, Norway, Australia and New Zealand. It has also been adopted in some form by a number of local regulatory bodies in the US, e.g. the public utility commissions in Colorado [37]:

> The utility shall include the concept of prudent avoidance with respect to planning, siting, construction, and operation of transmission facilities. Prudent avoidance shall mean the striking of a reasonable balance between the potential health effects of exposure to magnetic fields and the cost of impacts of mitigation of such exposure, by taking steps to reduce the exposure at reasonable or modest cost. Such steps might include, but are not limited to 1) design alternatives considering the spatial arrangement of phasing of conductors; 2) routing lines to limit exposures to areas of concentrated population and group facilities such as schools and hospitals; 3) installing higher structures; 4) widening right of way corridors; and 5) burial of lines.

Some health departments have also adopted policies or published informational literature that recommends prudent avoidance as a policy tool. They assessed the scientific literature on electromagnetic fields (EMFs) and concluded that adverse health effects from exposure to EMFs have not been established. Most of them agreed that there is some evidence that EMF exposure may pose a risk to health, and they suggested a cautious approach when building new electrical facilities, homes and schools (including kindergartens and child care structures) near existing electrical facilities such as power lines and substations. However, all have rejected imposing arbitrary low numeric EMF exposure levels since these are not supported by the scientific literature.

In the context of this chapter it is of great interest what "prudence" means in this approach. Coming from Aristotelian ethics, prudence denotes a pragmatic approach to what should be done. In this case, prudence consists first of balancing two frequently diverging ideas: the idea of precaution and the idea of making use of the technical advance even in case of unclear risk. Secondly, prudence is asked for in the concrete design of the technical facilities – alternatives are searched for that which might be similarly useful, but would create lower radiation loads for people concerned.

15.4.3
Interim Conclusions (2)

Both the approaches described above do not have an explicit ethical foundation. While the precautionary principle might be understood as an attempt to make Jonas' work (which stands in the Kantian tradition) more operable by involving assessment procedures concerning the knowledge available about the unclear risk under consideration, the prudent avoidance principle is obviously motivated by the Aristotelian approach. With respect to the outcome and the consequences, both approaches converge strongly. They lower the expectations with respect to ethical guidance and transform them into questions of "prudent" knowledge assessment which then shall guide the search for adequate actions and decisions. Consequently, they aim at solving the paradoxical challenge to enable assessments where assessments do not seem to be possible (*cf.* [37]). This challenge, however, can only be met by a procedural approach where the involvement of participatory processes seems plausible [33].

15.5
Conclusions

Strategies for dealing with unclear risk require careful normative reflection [4]. Situations involving unclear risk imply the absence of a "standard situation" in moral, in epistemic and in risk respects [11]. Ethical analysis is, therefore, required (Section 15.2) but cannot offer decisive guidance to respective decision-making situations (Section 15.3). Operable approaches (Section 15.4) transform the problem under consideration into procedures of assessing the knowledge available concerning the unclear risk. This transformation changes the guidance which could be expected to come from ethical reflection. Ethical investigation will not be able to give "categorical" guidance about what should be done and what should not be done.

Instead, ethical reflection is needed to uncover and to shed light on the normative premises of the options at hand, on the criteria of decision making as well as on the criteria of comparing different types of unclear risk of weighing unclear risks against (perhaps clearer) benefits. Such an ethical "enlightenment" is a necessary precondition for deliberative procedures within which society could identify adequate levels of protection, threshold values or action strategies. Questions of the acceptability and comparability of risks, the advisability of weighing up risks against opportunities, and the rationality of action under uncertainty are, without doubt, of great importance [38].

The possibility of ethical expertise can, therefore, be demonstrated methodically, but leads to a modest appreciation of ethics' possibilities. In particular, ethical expertise cannot give answers to the question of what should be done and how one should act. Society is left to itself, as far as planning the future and setting the course are concerned; ethics cannot take these aids to orientation from society's shoulders. Ethics provides analytical tools to uncover normative premises and to provide guidance how to assess the situation – ethical reflection informs societal and political decision making with regard to the normative issues involved, but cannot replace them. Decision making in cases of unclear risk is, therefore, bound to procedures of knowledge assessment, including appropriate discursive procedures, which needed to be informed by ethical analysis in order to work transparently in a normative respect.

References

1. Stern, P.C. and Fineberg, H.V. (eds) (1996) *Understanding Risk: Informing Decisions in a Democratic Society*, National Academy Press, Washington, DC.
2. Shrader-Frechette, K.S. (1991) *Risk Analysis and Scientific Method*, Reidel, Dordrecht.
3. International Program on Chemical Safety (2004) *Risk Assessment Terminology – Part 1 and Part 2*, World Health Organization, Geneva.
4. von Schomberg, R. (2005) The precautionary principle and its normative challenges. In *The Precautionary Principle and Public Policy Decision Making*, (eds E. Fisher, J. Jones and R. von Schomberg), Edward Elgar, Cheltenham, pp. 141–165.
5. Schmid, G., Brune, H., Ernst, H., Grünwald, W., Grunwald, A., Hofmann, H., Janich, P., Krug, H., Mayohr, M., Rathgeber, W., Simon, B., Vogel, V. and Wyrwa, D. (2006) *Nanotechnology – Perspectives and Assessment*, Springer, Berlin.
6. Grunwald, A. (2007) Nanoparticles: risk management and the precautionary principle. In *Ethical, Legal and Social Implications of Nanotechnology*, (ed. F. Jotterand), in press, Springer, Berlin.
7. Group, E.T.C. (2003) *The Big Down. Atomtech: Technologies Converging at the Nanoscale*, ETC Group, Ottawa. [http://www. etcgroup.org] [Retrieved: 2.6.2007].
8. Luhmann, N. (1989) *Paradigm Lost*, Suhrkamp, Frankfurt.
9. Gethmann, C.F. and Sander, T. (1999) Rechtfertigungsdiskurse. In *Ethik in der Technikgestaltung. Praktische Relevanz und Legitimation*, (eds A. Grunwald and S. Saupe), Springer, Berlin, pp. 117–151.
10. Grunwald, A. (2000) Against over-estimating the role of ethics in technology development, *Science and Engineering Ethics*, **6**, 181–196.
11. Grunwald, A. (2003) Methodical reconstruction of ethical advises. In *Expertise and Its Interfaces*, (eds G. Bechmann and I. Hronszky), Edition Sigma, Berlin, pp. 103–124.
12. Brown, N., Rappert, B. and Webster, A. (eds) (2000) *Contested Futures. A sociology of prospective techno-science*, Ashgate Publishing, Burlington.
13. Skorupinski, B. and Ott, K. (2000) *Ethik und Technikfolgenabschätzung*, ETH Zürich, Zürich.
14. Decker, M. and Grunwald, A. (2001) Rational Technology Assessment as Interdisciplinary Research. In *Implementation and Limits of Interdisciplinarity in European Technology Assessment*, (ed. M. Decker), Springer, Berlin, pp. 33–60.
15. Wardak, A. (2003) *Nanotechnology & Regulation. A Case Study Using the Toxic Substance Control Act (TSCA)*, Paper 2003-6 Woodrow Wilson International Center, Washington, DC.
16. Gannon, F. (2003) Nano-nonsense, *EMBO reports* **4**, 1007.
17. Ball, P. (2003) *Nanoethics and the Purpose of New Technologies*, Lecture at the Royal Society for Arts, London [www.whitebottom.com/philipball/docs/Nanoethics.doc] [Retrieved: 02.07.2007].
18. Gee, D. and Greenberg, M. (2002) Asbestos: from 'magic' to malevolent mineral. In *The Precautionary Principle in the 20th Century. Late Lessons from Early Warnings*, (eds P. Harremoes, D. Gee, M. MacGarvin, A. Stirling, J. Keys, B. Wynne and S. Guedes Vaz), Sage, London, pp. 49–63.
19. Harremoes, P., Gee, D., MacGarvin, M., Stirling, A., Keys, J., Wynne, B. and Guedes Vaz, S. (eds) (2002) *The Precautionary Principle in the 20th Century. Late Lessons from Early Warnings*, Sage, London.

20 Rescher, N. (1983) *Risk. A Philosophical Introduction to the Theory of Risk Evaluation and Management*, University Press of America, Lanham, MD.

21 Keeney, R.L. (1984) Ethics, decision analysis, and public theory, *Risk Analysis*, **4**, 117–129.

22 Gethmann, C.F. and Kamp, G. (2000) Gradierung und Diskontierung von Verbindlichkeiten bei der Langzeitverpflichtung. In *Die Zukunft des Wissens*, (ed. J. Mittelstraß), Akademie-Verlag, Berlin, pp. 281–295.

23 Birnbacher, D. (1991) Ethische Dimensionen bei der Bewertung technischer Risiken. In *Technikverantwortung*, (eds H. Lenk, and M. Maring), Campus, Frankfurt, pp. 131–146.

24 Caplan, P. (2000) *Risk Revisited*, Pluto Press, London.

25 Nida-Rümelin, J. (1996) Ethik des Risikos. In *Angewandte Ethik. Die Bereichsethiken und ihre theoretische Fundierung*, (ed. J. Nida-Rümelin), Alfred Kröner Verlag, Stuttgart, pp. 806–831.

26 Gethmann, C.F. and Mittelstraß, J. (1992) Umweltstandards, *Gaia – Ecological Perspectives for Science and Society*, **1**, 16–25.

27 Grunwald, A. (2005) Zur Rolle von Akzeptanz und Akzeptabilität von Technik bei der Bewältigung von Technikkonflikten, *Technikfolgenabschätzung – Theorie und Praxis*, **14**, 54–60.

28 Jonas, H. (1979/1984) *Das Prinzip Verantwortung [The Imperative of Responsibility]*, Suhrkamp, Frankfurt/M.

29 Grunwald, A. and Kopfmüller, J. (2006) *Nachhaltigkeit*, Campus, Frankfurt.

30 Dupuy, J.-P. and Grinbaum, A. (2004) Living with uncertainty: toward the ongoing normative assessment of nanotechnology, *Techné*, **8**, 4–25.

31 Joy, B. (2000) Why the future does not need us, *Wired Magazine*, April, 238–263.

32 Harris, J. (1975) The survival lottery, *Philosophy*, **50**, 81–87.

33 Pereira, A.G., von Schomberg, R. and Funtowicz, S. (2007) Foresight knowledge assessment, *International Journal of Foresight and Innovation Policy*, **3**, 53–75.

34 Morgan, M.G. (1992) Prudent avoidance, *Public Utilities Fortnightly*, March 15.

35 Dekay, M., Small, M., Fischbeck, P., Farrow, R.S., Cullen, A., Kadane, J., Lave, L., Morgan, M.G. and Takemura, K. (2002) Risk-based decision analysis in support of precautionary policies, *Journal of Risk Research*, **5**, 391–417.

36 Nair, I., Morgan, M.G., Florig, H.K. (1989) *Biologic Effects of Power Frequency Electric and Magnetic Fields (OTA-BP-E-53)*, Office of Technology Assessment, Washington, DC.

37 Colorado Public Commission (1992) *Statement of Adoption in the Matter of the Rules for Electric Utilities of the Colorado Public Utilities Commission. Code of Colorado Regulation-723-3 Concerning Electric and Magnetic Fields*, Colorado Public Commission, Denver, CO.

38 Grunwald, A. (2004) The case of nanobiotechnology. Towards a prospective risk assessment, *EMBO Reports 5 (Special Issue)*, 32–36.

V
Practical Implications

16
Lessons Learned: Recommendations for Communicating Conflicting Evidence for Risk Characterization

Peter Wiedemann, Franziska Börner, and Holger Schütz

16.1
Introduction

As the previous chapters have shown, risk characterization is concerned with the decision-making-oriented presentation and summary of the insights gained from a multistep assessment process regarding the hazardousness of an agent to human health. In theory, this forms the basis of risk management, and serves the needs and interests of decision makers and stakeholders.

However, the complexity of the individual risk characterization processes as well as the variety of approaches and standards used by risk assessors and regulatory bodies make the dissemination of the gained insights a real challenge. Additionally, regulatory decision making is also increasingly influenced by public reactions regarding real or putative hazards, as Evi Vogel and Ginevra Delfini have stressed in Chapter 3. These developments show that it is essential for all actors in the risk community to communicate their assessments not only to the small circle of policy makers, but also to concerned parties and the broader public. Nearly inevitably, different views, controversy and conflict come into play.

In other words, risk assessment has become a battlefield, as Paul Slovic argued [1]. It seems that his evaluation is still valid, especially when we take a look at the controversies around unclear risks caused by conflicting evidence. Apart from electromagnetic fields (EMFs), examples of unclear risks causing controversies and risk discussions arise not only from the fields of emerging technologies like biotechnology and nanotechnology, but also in more established areas, e.g. with respect to the suspected hazardousness of already existing chemicals.

In the past, two different (however, we argue) complementary approaches have been developed that aim to provide solutions to this battlefield.

The first approach refers to the concept of good risk communication, first elaborated as *Seven Cardinal Rules of Risk Communication* in a US Environmental Protection Agency (EPA) brochure in the late 1980s [2] and recently adapted in a booklet entitled *Establishing a Dialogue on Risks from Electromagnetic Fields* by the World Health Organization in order to support decision makers [3]. The basic idea

The Role of Evidence in Risk Characterization: Making Sense of Conflicting Data.
Edited by Peter M. Wiedemann and Holger Schütz
Copyright © 2008 WILEY-VCH Verlag GmbH & Co. KGaA, Weinheim
ISBN: 978-3-527-32048-6

here is to solve risk controversies by dialogue, partnership and participatory decision making.

The second approach is focusing on better science. A prominent example here is the influential book on phantom risk edited by Foster, Bernstein and Huber in 1993 [4]. The book explores a special class of risks, i.e. risks that cause remarkable public attention, but are unproven or even unprovable. According to Foster *et al.*, phantom risks arise from errors in science, from ambiguity as well as from observational biases of patients or doctors. For these authors, the key to overcome these problems is a better and more careful appraisal of scientific evidence.

Careful appraisal of scientific evidence is a demanding task. It starts with the assessment of the quality and the explanatory power of single studies. Furthermore, a critical issue in the summing up review process consists in integrating conflicting study outcomes. Finally, various types of research, e.g. mechanistic studies and animal research, have to be synthesized in order to arrive to the conclusion: does the overall scientific picture point to human health risk or not?

The contributions in our book show that the three research pillars of risk assessment, i.e. epidemiology, animal research and genotoxicological or mechanistic studies, have developed their own individual frameworks and standards in order to ensure sound science for better risk assessment. The procedures for characterizing evidence covered in this book have developed criteria for reviewing and weighting scientific evidence that are published and promoted by various scientific societies in the area of medicine and public health.

However, the situation is sometimes more difficult and the problems often lie within the details. Obviously, experts agree in principle that the quality standards commonly used in science must be applied in evaluating the results of scientific findings. However, opinions differ in part dramatically regarding specific standard details. For example, some scientists argue that statistical significance (1 or 5% significance level) should not be a knockout criterion in the evaluation of a study [5]. They feel that even statistically nonsignificant trends that become apparent in the data provide important information for the interpretation of the results, especially if these trends can be found for different parameters or across different studies. Whether such a shift in criteria for acceptance or rejection of study results is justified in risk assessment can lead to controversial debates among scientists.

This example points to the necessity of further methodological discussion regarding standardization within and between different scientific expert communities. However, the remaining of the chapter does not focus on these methodological discussions. Instead, we will highlight the communication issues that are related to the appraisal of scientific evidence for assessing human heath risks. Here, our goal consists in outlining recommendations that focus on how to take conflicting evidence into account and how to explain unclear risks to audiences outside the specific scientific communities.

These following recommendations are inspired by four principles originally proposed by the US EPA in its handbook on risk characterization [6]: *transparency*, *consistency*, *clarity* and *reasonableness*. From this list of EPA principles, only consistency has no direct relevance for risk communication, all others can be easily applied with respect to communicating scientific evidence. However, further criteria seem to be necessary. Thus, we extend the EPA list by including three other principles: *prudence, impartiality* and *responsibility*.

16.2
Guiding Principles in Risk Communication

16.2.1
Prudence

We understand prudence as a basic scientific virtue that covers both a reflective and careful judgment as well as a tendency to consider caution as a guiding principle of action taking, i.e. for risk management.

16.2.1.1 Assess the Underlying Problem

Prudence is based on a careful analysis of the situational circumstances. Risk communication should be based on a comprehensive consideration of the communication situation, including the diagnosis of the underlying communication problem. At least two issues have to be distinguished: comprehension problems and conflicts about risk. There is no doubt that risk comprehension is a very important objective for risk communication. However, the attitude to view all kinds of risk communication problems as comprehension problems leads to the mistaken assumption that "educating" the public will help solve any risk issue, including conflicts about risk. If in fact the risk problem at hand is a problem of comprehension, an education centered approach to risk communication is appropriate. If the risk problem, however, is actually a problem of conflict, education about risk is not sufficient, and insisting exclusively on an education approach might be even obstructive for conflict solving. Most often risk conflicts are based on a clash of interests or values and do not result from comprehension deficits of one of the parties involved in the conflict. The key to conflict solution – or at least mitigation – is negotiation. Nevertheless, informing about the available scientific evidence regarding a risk can be a valuable part of a negotiation process because it might help to separate scientific facts from values.

As for this, the analysis of how much of the risk controversy is based on differences resulting from clashing interests or conflicts and how much it depends on actual lack of risk knowledge is a crucial diagnosis.

16.2.1.2 Both Content and Process do Matter

Despite these cautions to regard risk communication only as education about risk, ignoring the information about risk would be equally misguiding. Suggestions such as "The primary goal of risk communication is to establish high levels of trust and credibility" ([7], p. 4) sometimes seem to overestimate the importance of process variables like trust and confidence, and to underestimate the importance of the content of risk communication, i.e. the messages about the risk itself and the amount and quality of scientific evidence. Audiences should be informed about the type of evidence and its explanatory power for a human health risk in order to make informed decisions. For instance, they should be informed about the value and effectiveness of special study types, and whether or not a selected endpoint is relevant for risk assessment. Thus, more efforts should be made to explain the underlying framework of evidence appraisal.

16.2.2
Transparency

Transparency is an additional key factor for communicating appropriately on the results of weighting and appraising evidence. It highlights explicit underlying assumptions, contexts and premises, and indicates how experts have reached their results.

16.2.2.1 Make Your Expertise Transparent
Each scientist involved in a risk assessment process should provide basic information regarding his or her qualifications. A risk assessment is based on integrating various fields of scientific expertise, e.g. the EMF risk field covers dosimetry, biophysics, biology, animal research, epidemiology, as well as genotoxicology and molecular medicine, etc. Thus, it is very important to reveal the individual scientific competency of the involved experts. Furthermore, the expert should disclose his or her level of experience and familiarity with experimental studies and empirical investigative methods. Thorough experience and knowledge are crucial for the critical evaluation of scientific evidence. For better transparency, expert profiles could be generated and made accessible. Such profiles should not only disclose each researcher's scientific core expertise, but also include their fields of professional work and the scientific bodies the researcher is associated with.

16.2.2.2 Describe the Context of Your Work and the Process of Arriving at the Conclusion
Context and process do matter. For instance, it is important to communicate how the reviewing process for scientific evidence was organized, who was involved and whether consultations or cooperations with other experts have been done. Of special importance is also whether the spectrum of different scientific positions was represented by the involved experts. Furthermore, it is interesting to know, whether independent researchers participated as reviewers in the assessment process.

16.2.2.3 Reveal your Evaluation Framework
In the final section of his famous article on causality criteria, Sir Bradford Hill [8] mentions that in real life we may surely look at the consequences of our decision when we evaluate scientific evidence. Depending on what is at stake, Hill suggests that in real life different standards for decision making and action taking are used. In some cases, weak evidence is already enough to take action; in other cases, strong evidence is necessary before actions should be taken. Therefore, it is of the utmost

importance to disclose the standards used in weighing evidence. For example, if a scientist adopts a precautionary perspective framework for risk assessment, they may focus more strongly on positive than on negative results and may put less weight on the methodological rigor of the respective studies. Others might first look at the methodological quality of a study and only if it is sufficient take the study results into consideration. The evaluation framework, i.e. the guidelines and principles on which the assessment of evidence is based, need to be made known.

16.2.2.4 Describe the Rules that You Use for Evaluating the Weight of Evidence

In order to fully understand how a risk assessor comes to a final judgment, it is necessary to know how and by which rules the overall scientific picture is created. With regard to risk assessment, the International Agency for Research on Cancer (IARC; see Cogliano *et al.* in Chapter 9) has played here a pioneering role. Such rules are not only an indispensable precondition for a consistent appraisal of evidence, they are also helpful for outsiders to understand the overall scientific picture risk characterization is based on. Such rules provide a supporting framework to deal with inconsistent findings, e.g. among animal studies and epidemiology, and help to bring forth a substantiated risk judgment.

16.2.3
Impartiality

Impartiality means open-minded thinking that is free of bias, prejudice or certain outcome preferences based on nonscientific motives.

16.2.3.1 Give the Pros and Cons of Your Assessment

One of the critical issues in summing up evidence is the confirmation bias, i.e. the propensity to interpret studies in a way that confirms one's opinion and downplays studies which contradict one's own beliefs. A good way to demonstrate fairness is to address explicitly both sides: the arguments speaking for the existence of a risk and the arguments against it. This information allows one not only to compare the strength of the competing arguments, but also provides a solid basis for one's own conclusion.

16.2.3.2 Depict the Remaining Uncertainties but Do Not Forget to Point Out the Evidence Already Available

The characterization of the remaining uncertainties in risk characterization in a way that enables the audience to understand the strength as well as the weakness of the available evidence is crucial. Research has shown (see Wiedemann *et al.* in Chapter 14) that contrary to the assumption that revealing uncertainties will strengthen the public's trust into the risk assessors, it appears that the disclosure of uncertainties might also have negative effects. Empirical research indicates that nonexperts have difficulties to differentiate among different types of uncertainty and tend to draw incorrect conclusions in their own risk judgments. There is no research yet on how to address these fundamental comprehension problems of lay people on the nature of uncertainty. However, we suggest with all caution that uncertainty information should be accompanied by a clear description of the available evidence so far and to portray both sides of the evidence, i.e. what is known and what is not known.

16.2.4
Reasonableness

Communicating scientific evidence should be based on sound critical thinking, i.e. reasonableness.

16.2.4.1 Explain the Process of Evaluating Evidence
Nonexperts are unfamiliar with the process of hazard assessment. They often do not have an idea of the complexity of the processes and the errors that can result. Therefore, communication about hazard assessment should at least roughly inform its audience about the main principles of hazard identification. The goal is to establish an understanding that a general conclusion about the existence of a hazard may not be drawn from a single study alone. Instead, evidence from more than one research area has to be taken into account.

16.2.4.2 Explain the Relevance of the Endpoints for Evaluating Human Health Risks
There is a widespread agreement among risk assessors that a distinction must be made between biological effects and detrimental health effects. This distinction is indeed central to risk assessment, since any effect is relevant for risk assessment only insofar as this effect can be considered to be damaging to health. For instance, is a shortening of response times in cognitive reaction tasks under radiofrequency (RF) EMF exposure an indicator of health risks? Foremost, it indicates an improvement in cognitive performance and it is hard to see how it should be indicative of a detrimental health effect. Therefore, it matters to inform people about this distinction.

16.2.4.3 Put the Available Evidence in Perspective
The number of available studies with respect to a certain endpoint is essential and has to be made public. It makes a great difference in the evidence evaluation process whether there is only one study or at least one additional replication study, or even a range of studies available. With only one available study considerable uncertainties remain. However, if many studies are available conclusive judgments are possible, depending on the homogeneity of the findings.

Additionally, information about the number of available studies also allows comparative judgments. Those evidence comparisons can be made with respect to the sheer number of available studies as well as the amount of available evidence for an adverse health effect. Furthermore, comparisons can be made between different endpoints (where is there more need for further research, with respect to brain cancer or leukemia?) and between different agents. For the latter case, a good example is the comparison of the available evidence regarding carcinogenicity from ELF versus RF EMF exposure.

Primarily, comparisons serve to establish understanding. Hence, informing people about evidence comparisons should help them to develop a more realistic picture about the available evidence and to make better judgments.

16.2.4.4 Support Accessibility of Critical Information

A visual depiction is often helpful to grasp complex information. With regard to risk assessment, this can be done by presenting the three core elements of risk characterization, i.e. the evidence basis, the pro and con arguments, and the conclusions together with the remaining uncertainties, in a visual way, e.g. as an evidence map.

16.2.4.5 Assess the Potential Risk

For a proper evaluation of the seriousness of a potential risk, it is helpful to estimate how big that risk would be, given that the hazard actually exists. However, informing the public about such a conditional risk estimate is a precarious task, as this information might be misinterpreted to suggest that the hazard actually exists. Nevertheless, only risk estimates – and not hazard information alone – allow an evaluation of the public health impact of an agent.

16.2.4.6 Put the Potential Risk in Perspective

Comparisons play a crucial role in risk communication. They serve to clarify the nature, magnitude and relative significance of a risk. It should be noted, however, that a better understanding of a risk does not imply that the acceptance of a risk will change.

16.2.5
Clarity

Clarity refers to a use of language that supports comprehension, retention and understanding. It serves knowledge building.

16.2.5.1 Give No More Information than Necessary

A basic misapprehension is to believe that the more information is given, the more clarity will be accomplished. However, more information does not automatically imply more knowledge. As knowledge requires critical reflection of information, people need advice on how to interpret the given information. To support knowledge building, people need to focus on the critical core information as too many details and information may confuse them and cause irritation. Therefore, the message should be careful designed around the main arguments.

16.2.5.2 Be Aware of Your Language

Communicators should make an effort to help people to clearly distinguish between risk and hazard. For example, if lay persons hear the term relative risk, they will likely interpret it as a measure of risk and not as a measure of association which is used as one of several pieces of evidence to determine whether an agent is hazardous or not.

Unfortunately, especially in the field of epidemiology, risk and hazards are not easily distinguishable for lay people, because an indicator for a statistical association (relative risk) is used to describe the evidence for a hazard. For an inexperienced reader, relative risk might suggest that the study is reporting a risk, while it is actually reporting the evidence regarding a hazard.

16.2.5.3 Test the Perceptions of your Communication Formats

Sometimes, the choice of a certain communication format for presenting the results of evidence appraisal is based just on intuition. For instance, the German Radiation Protection Commission [(Strahlenschutzkommission) (SSK)] has used a three-level approach for evidence characterization. The SSK differentiates among established evidence, justified scientific suspicion and scientific indication (see Leitgeb in Chapter 11). Contrary to the logic of the SSK, who consider suspicion ranking higher than indication in terms of strength of evidence, empirical research indicates that lay people tend to believe that a scientific indication is stronger in evidence than a scientific suspicion. Thus, the core message of the SSK is distorted by diverging understanding of these two labels. This example is not a singular case; counterintuitive effects of risk communication are not an exception (see Wiedemann *et al.* in Chapter 14). Therefore, we strongly recommend evaluating the effects of risk communication activities in a systematic way before implementation.

16.2.6
Responsibility

Any risk assessment serves an ultimate goal: to create a solid knowledge base for the sound judgment whether an agent poses a risk to humans or not. In this context, responsibility refers to the moral obligation to deliver a sound and logical risk characterization in order to provide the best possible foundation for risk management, i.e. to protect people from real harm, on the one hand, and to avoid false warnings, on the other.

16.2.6.1 How Much Evidence is Evidence Enough for Taking Action?

Knowledge gaps, and inconsistent and conflicting data result in an inconclusive risk characterization. As a result, risk assessors are not able to give a definitive judgment whether an agent is harmful or not. In such cases risk managers are confronted with a difficult decision.

A key issue is the standard for taking action, i.e. whether assessors should suggest protective or precautionary measures in order to prevent possible health damage. On the one hand, scientific scrutiny is required to protect against premature and inappropriate judgment; on the other hand, cautious consideration of the evidence is needed in order to detect weak signals of emerging risks. A balanced judgment has to be achieved. However, this balance will – at least in part – depend on subjective standards. For some risk assessors, a single study may be a sufficient ground for suggesting precautionary measures, others will when confronted with conflicting evidence tend to recommend actions and still others will wait for further clarifying studies before they advise any actions.

These standards inevitably include a subjective element and any attempt to substitute them by an objective criterion will be facing insurmountable obstacles. To us the only feasible solution seems to be to follow the principles of risk communication outlined in this chapter in order to find a negotiated solution. Engaged in a fair dialogue with stakeholders and concerned parties, risk assessors should try to find an acceptable standard for action.

References

1 Slovic, P. (1999) Trust, emotion, sex, politics, and science: surveying the risk-assessment battlefield, *Risk Analysis*, **19**, 689–701.
2 Covello, V.T. and Allen, F.W. (1988) *Seven Cardinal Rules of Risk Communication (OPA-87-020)*, US Environmental Protection Agency, Washington, DC.
3 WHO (2002) *Establishing a Dialogue on Risks from Electromagnetic Fields*, World Health Organization, Geneva.
4 Foster, K.R., Bernstein, D.E. and Huber, P.W. (eds) (1993) *Phantom Risk: Scientific Inference and the Law*, MIT Press, Cambridge, MA.
5 Schütz, H. and Wiedemann, P.M. (2005) How to deal with dissent among experts. Risk evaluation of EMF in a scientific dialogue, *Journal of Risk Research*, **8**, 531–545.
6 EPA (2000) *Risk Characterization Handbook (EPA 100-B-00-002)*, US Environmental Protection Agency, Washington, DC.
7 MOA (2004) *Risk Communication Handbook – "Working with the Community"*, Mobile Operator Association, London [http://www.mobilemastinfo.com/planning/Risk_Communication_Handbookv2.pdf] [Retrieved: 20.03.2006].
8 Hill, A.B. (1965) The environment and disease: association or causation?, *Proceedings of the Royal Society of Medicine*, **58**, 295–300.

Index

a

acceptability 189
– unclear risk 189
activist groups 29
alcohol 45
ambient monitoring 84
ambiguity 169ff., 172
Ames test 42
analogy 104, 136, 143
aneugenic action 41
animal experiments 102
animal studies 4, 55
approach 20
– decisionist 21
– deliberative 21
– scientific 22
– technocratic 20
argument anatomy 139
argumentation theory 153
arguments 147, 153
– causal 147
– con 153
– pro 153
association 103, 138
– strength 73

b

base rate information 169
base station(s) 32, 56
Bayesian analysis 18
Bayes theorem 139–140
beliefs 174–175, 178, 179
– prior 175, 178
– uncertain 174, 179
bias 70, 74–76, 97, 102, 135, 139, 175
– negativity 175
– participation 70, 74
– recall 70
– reporting 74
– selection 70, 74, 76
bioassays 102, 104
biologic effects 122
biological plausibility 76, 104
biomonitoring 83
blinded design 58
brain cancer 145
– childhood 145
brain tumor(s) 69, 72, 117, 156ff.

c

cage controls 58
cancer 101
– epidemiology 154ff.
carcinogen(s) 101
– chemical 101
– identification 101
carcinogenicity 6, 103ff., 112, 146
cardiovascular effects 17
case history 81, 83
case report 87, 88
causal model 137
causal relationship 6–7
causality 102ff., 138, 141, 145
causation 134, 142
cause–effect relationships 188
cell cycle 37
certainty 6–7
– degree of 6
chance 135
childhood leukemia 2, 5, 16–17, 19, 67ff., 123–124
chromosomal aberrations (CAs) 37ff., 39, 43
clarity 211
clastogenic action 41
clinical case studies 5

clinical environmental medicine 81
clinical expertise 93
coherence 104, 136, 142
COMET assay 38
communication formats 212
competency 208
– scientific 208
confidence 30ff.
conflicts of interests 108
confounding 71, 75, 97, 102, 135, 139
consequentialism 192
consequentialist approach 191, 195
consistency 74, 135, 192
– principle of pragmatic 192
context 174ff., 179
critical reflection 211
critical thinking 210

d
decision analysis 131
decision making 2–3, 32–33, 188–189
– regulatory 3
– uncertainty 2
deontological advice 195
diagnostic protocol 86
disagreement 165
– among experts 165
DNA 37ff.
dose–effect relationship 46
dose–response 135
– assessment 1, 107, 164
– relationship 60, 73

e
effect(s) 16, 55–56, 73, 84, 113, 210
– acute 16
– biological 210
– biophysical 56
– chronic 16
– genotoxic 113
– health 210
– monitoring 84
– thermal 55
– threshold 73
electromagnetic field(s) xiii, xv, 13, 50, 99, 111, 121, 152, 199
– radiofrequency (RF) xiii, 126
electromagnetic hypersensitivity 123–124
electromagnetic interference 128
electrosensitivity 8, 84, 88
ELF, see extremely low frequency
EMF, see electromagnetic fields
emission minimization 127
endpoint(s) 47ff., 210

– relevance of 210
environmental attribution(s) 82, 85
environmental clinical case studies 81
environmental contamination 84
environmental medicine 81
epidemiological studies 101ff.
epidemiology xiii, 4, 6–7, 67, 155
– cancer 155
ethical approaches 191
– unclear risk 191
ethical guidance 185ff.
ethics 8, 186, 191
– consequential 191
evaluation 98
– framework 208
evidence 3ff., 5, 25, 28–29, 33, 95, 205, 207, 210
– best available 95
– carcinogenicity 102ff.
– categories 116
– characterization 3ff., 5, 47, 158
– comparisons 210
– conflicting 205
– evaluation 210
– evidence 104, 136
– levels 5, 95, 102ff.
– map(s) 7, 151ff., 211
– rating 112
– science-based 3, 28
– scientific 33, 207
– society-based 29
evidence-based clinical medicine 94
evidence-based healthcare 94
evidence-based medicine 5, 84, 93
experimental animals 104ff.
experimentation 143
experiments 46
– independent 46
expert 26, 28, 33
– committees 28, 33
– judgment 26
– self-appointed 28
expert dissent 1
expertise 93, 208
– clinical 93
– scientific 208
exposure 60
– assessment 1, 23, 76, 83, 107, 119, 164
– levels 57, 113
– life-long 60
– limits xv, 2, 57
– long-term 60
– misclassification 103
– response relationship 103

– systems 57, 58
extremely low frequency
– magnetic fields xiii, 2, 6, 13ff., 67, 124, 126

f
fact sheets 19
fibroblasts 43

g
gene expression 56
genotoxic agents 37ff.
genotoxicity 4, 46
– testing xiii
good laboratory practice (GLP) 27, 43, 45
grading system 5

h
hazard 107
– characterization 23
– future 108
– identication 23, 67, 107–108, 164
– potential 1
health complaints 81
health-relevant effects 122
healthy cohort effect 71
hedging phrases 171
heuristics of fear 193–194
Hill's criteria 72, 77, 103
Hill's aspects 134ff.
homogeneity 135
human peripheral blood lymphocytes (HPLs) 43, 38

i
impartial evaluations 108
impartiality 209
in vitro studies 43
inadequate data 125
incidence rate 68
individual risk 116
information campaigns 32
informed decisions 208
innumeracy 173
interdisciplinary clinical diagnostics 85
interphone study 157
intuitive toxicology 167–168, 179
involvement 32
– public 32

j
judgment(s) 26, 107, 148, 165
– expert 26
– informed 165
– professional 148

– scientific 107
justification 187
justified scientific suspicion 125

k
knowledge 188
– gaps 212
– lack of 188

l
lack of data 117
level of evidence 95, 102, 126

m
mammals 44
mechanistic plausibility 136
media 1, 31
melatonin 77, 84, 123
meta-analysis 73–74, 118
metabolic activation 44
micronuclei 41ff., 43
misclassification 70, 76
mobile phone(s) 29, 56, 69, 113, 117, 155ff., 177
– pact 33
monitoring 84–86
– ambient 84–86
– bio 86
– effect 84
– susceptibility 84
morals 186, 187
multiple chemical sensitivity 88
mutagenic 37

n
nonspecificity 141

o
odds ratio 69, 155–156
on-site inspection 84

p
pedigree 137–138
phantom risk 206
physical interactions 122
physiologic reactions 122
plausibility 143
policy drafters 3, 25ff., 30
policy making 2–3, 25ff., 30–31
politicians 30
potential risk 211
power frequency magnetic fields 131ff.
power lines 124
pragmatic consistency 192

precautionary approaches   189
precautionary measures   176ff., 177, 212
precautionary principle   xv, xvi, 2, 8, 126, 197
premature chromosome condensation   39
probability description   173
projected time   194
prudent avoidance   8, 198
public participation   34
publication bias   117

r

radiation   29, 45, 50, 55, 73, 111, 127
– dose   45
– ionizing   45, 73
– nonionizing   29, 50, 55, 111
– protection   127
– ultraviolet   127
radiation protection   121
radon   69
randomization   95, 97
randomized control trial(s)   95, 104
rating evidence   116
reasonableness   210
regulations   126
relative likelihood   134
relative risk   68, 71, 73, 76, 157, 168–169, 211
relevance to health   113
replication studies   59
responsibility   193–194, 212
– imperative of   193–194
review   15
– of the literature   15
risk appraisals   7
risk assessment(s)   xiii, 1, 14, 17, 22, 99, 117, 121, 163–164, 205ff.
– uncertainty   163ff., 164
risk characterization   1, 10, 23, 205, 206ff.
risk communication   xiii, 1, 2–3, 7, 13–14, 19ff., 23, 29, 82, 163, 178–179, 205ff., 207
– evidence-based   179
– guiding principles   207
risk comprehension   207
risk controversy   207
risk difference   68
risk estimation   67
risk evaluation   131
risk knowledge   7, 207
risk management   1, 23
risk perception   7, 121, 166ff., 169–170
risk regulation   185ff.

s

scientific evidence   124

scientific indication   125
sensitivity   141
sham exposure   58
significance testing   158
sister chromatid exchanges   37ff., 41, 44
specific absorption rate   57
specificity   74, 136
static fields   17
statistical significance   134
strand breaks   37, 56
– double   37ff.
– single   37ff.
studies   6, 18, 37–38, 43–44, 68–70, 72, 95, 101–102, 105, 112, 124
– analytical   68
– animal   6
– case-control   68–69, 72, 95, 101
– cohort   68, 70, 72, 95, 101
– cross-sectional   68–69
– descriptive   68
– ecological   69
– epidemiological   101, 112, 124
– experimental   6, 112
– in humans   112
– *in vitro*   18, 37–38, 43
– *in vivo*   18, 44
– mechanistic   6, 102, 105
– toxicokinetic   102
study quality   118
susceptibility   83–84
– genetic   84

t

temporality   136, 142
tenants consent procedure   32
teratogenicity   113
threshold effect   73
tissue   57
– biological   57
toxicology   4, 37ff.
– genetic   4, 37
transparency   208
trust   176, 178

u

uncertain beliefs   174
uncertainty   22, 26–27, 29, 163, 166, 169, 171–173
– categories   172
– communication   166
– descriptions   171, 173
– information   166
– scientific   26–27, 29
unclear risks   163ff., 185ff., 205

v
visual aids   174
visual depiction   211

w
weighing   189, 195
– benefits against risks   189
– risks and benefits   195
weight-of-evidence   15, 19, 101ff., 107, 153, 209
willingness to certify   132ff.
worst-case scenarios   194